# CHROMATOGRAPHIC
# INTEGRATION METHODS

# RSC Chromatography Monographs

Series Editor: Roger M. Smith, *University of Technology, Loughborough, UK.*

Advisory Panel: J. C. Berridge, *Sandwich, UK,*
G. B. Cox, *Delaware, USA,* I. S. Lurie, *Virginia, USA,*
P. J. Schoenmakers, *Eindhoven, The Netherlands,*
C. F. Simpson, *London, UK,* G. G. Wallace, *Wollongong, Australia.*

This series is designed for the individual practising chromatographer, providing guidance and advice on a wide range of chromatographic techniques with the emphasis on important practical aspects of the subject.

### Supercritical Fluid Chromatography
Edited by Roger M. Smith, University of Technology,
Loughborough, UK.

### Chromatographic Integration Methods
by N. Dyson, Dyson Instruments Ltd.,
Houghton le Spring, UK.

*Forthcoming titles*

### Electrochemical Detection and Liquid Chromatography in the Biosciences
by C. A. Marsden, Queen's Medical Centre, Nottingham, UK, and D. Perrett, St. Bartholemew's Hospital, London, UK.

### Supercritical Fluid Extraction: Laboratory Techniques and Applications
by M. E. P. McNally, E. I. Du Pont De Nemours & Co., Wilmington, DE, USA, and Z. Otero Keil, Keil Associates, Newark, DE, USA.

*How to obtain future titles on publication*
A standing order plan is available for this series. A standing order will bring delivery of each new volume immediately upon publication, at a substantial discount price. For further information, please write to:
The Royal Society of Chemistry
Distribution Centre
Blackhorse Road
Letchworth
Herts. SG6 1HN

Telephone: Letchworth (0462)672555

RSC
CHROMATOGRAPHY
MONOGRAPHS

# *Chromatographic Integration Methods*

**Norman Dyson**
*Dyson Instruments Ltd.*
*Houghton le Spring, UK*

ROYAL
SOCIETY OF
CHEMISTRY

**British Library Cataloguing in Publication Data**
Dyson, N. (Norman)
  Chromatographic integration methods.
  1. Chromatography. Measurement. Techniques.
  I. Title    II. Royal Society of Chemistry    III. Series
  543.089

  ISBN 0-85186-587-9

Published by The Royal Society of Chemistry,
Thomas Graham House, Science Park, Cambridge CB4 4WF

Filmset by Bath Typesetting Ltd., Bath,
and printed by Athenaeum Press Ltd., Newcastle upon Tyne

# *Preface*

This book is about the measurement of chromatographic peaks. In particular, it describes and discusses the manual and electronic techniques used to make these measurements, and how to use integrators. The aim of the book is simply to help analysts extract more data from their chromatograms, and help them to understand how integrators work so that results are never accepted unquestioningly.

The book is written for those who use integrators most; the lab technicians and students, and the analysts who do the chromatography. It is especially for those who never got around to reading their integrator manuals. When all else fails, read the manual or this book.

There are no books on integrators (to the author's knowledge), and the subject of integration tends to merit a short chapter in books on quantitative chromatography. The best survey of integration so far is Papas's 1989 review (Chapter 1, ref. 65). Those readers who want to study the subject more fully will find it indispensable. This book starts at the detector and works outwards. Most of it is about the integrity of peak representation, different methods of measuring peak data and validation of the results.

The electronics of data acquisition are kept to a minimum, enough to point the interested in the right direction, but not enough to redesign the front end of an integrator.

Current methods of integration measure peaks according to simple rules developed for manual measurement. Perpendiculars and tangents were shown years ago to have only limited justification yet they remain in universal employment because there is nothing better to replace them at the present time.

The goal of the 'definitive integrator' which would measure peak areas accurately, precisely, and instantly was scheduled to develop in three broad stages:

(1) Manual methods involving measurement of peaks on strip chart recorder traces. From this developed:

(2) Low cost integrator and micro-computer techniques. These were meant to be short term measures, replaced at the earliest opportunity by:
(3) Computer curve fitting of peak shapes and deconvolution of overlapping peaks using techniques that did not require straight lines except in the few cases where they might be justified.

Manual measurement of chromatograms brought attention to what might be called the 'problems of integration': how to define peak boundaries (the limits of integration), the effects of asymmetry, separation of overlapping peaks, measurement of small peaks, signal to noise ratio, *etc.*

When the first commercial integrators were introduced, they eliminated the tedium of measuring peaks by hand, greatly speeded up the processing of analyses, and increased the precision and number of analyses which could be made in a working day.

What integrators did not do was improve the methods of integration. Manufacturers simply selected the best of the manual methods and built integrators to use them.

Curve fitting of experimental data to a good peak model was believed to be the correct way forward. There would be no need for perpendiculars, tangents, or the construction of artificial baselines. Unfortunately, finding a suitable peak model has so far proved to be impossibly difficult. Peak shape is a fluid thing influenced by many factors, such as solute polarity, quantity, the column, mobile phase composition, system dead volume, injection proficiency, detector geometry, electronics, heaters, flow controllers, other solutes, the solvent which the solutes are dissolved in, or any combination of these factors.

Defining peak shape requires all of these factors to be brought under control, but research has shown that even when this is done, and a good mathematical model created for a given peak, the model does not necessarily describe neighbouring peaks, nor does it describe the same peak a week later if the column has aged meanwhile.

It has been judged that there is simply not enough information in the output of a single channel detector such as an FID or UV detector to allow measurement of a whole chromatogram. Analysts cannot even be sure how many peaks there are in a group. Under these circumstances, contemporary integrators have limited powers of data interpretation. To paint the full picture, additional data must be supplied. It is possible that 'three dimensional detectors' such as the Diode Array, GC–MS or FTIR will supply the extra information to allow further progress. The 'next generation integrator' might well be based on the creation and processing of a three-dimensional field which is Gaussian in one plane and Lorentzian in the other, or approximately so. If so, it will worry the independent integrator manufacturers if they are denied information on interfacing to these complex detectors.

The purpose of this book is not 'to teach chromatographers how to use an integrator to measure peak areas correctly', because the principles on which integrators are based do not allow the accurate measurement of any peak unless it is completely resolved and stands on a flat, noise free baseline. The purpose is to describe the rules of integration as they stand, and the implications this has on peak measurements. Only by knowing these rules and working within their limits can the integrator be used to its best.

The moral of the story can be anticipated in advance: no matter how expensive the integrator or computer, it is no substitute for good chromatography by a capable and critical analyst. The quality of results is largely determined before the integrator is ever brought into use.

# *Contents*

| Chapter 1 | | **Theory of Peak Measurement** | **1** |
|---|---|---|---|
| | 1 | The Basic Measurements | 1 |
| | | Peak Area and Peak Height | 1 |
| | | Retention Time and Solute Identity | 3 |
| | 2 | Experimental Validation | 4 |
| | 3 | Results Validation | 8 |
| | 4 | Performance Measurement and Accountability | 8 |
| | 5 | Chromatographic Peaks and the Gaussian Function | 9 |
| | | The Function | 9 |
| | | Peak Height | 10 |
| | 6 | The Exponentially Modified Gaussian Function | 17 |
| | | Practical Application of the EMG Function | 19 |
| | | EMG Peak Shape Tests | 19 |
| | 7 | Statistical Moments of a Chromatographic Peak | 20 |
| | | Zeroeth Moment, $m_0$ | 20 |
| | | First Moment, $m_1$ | 20 |
| | | Second Moment, $m_2$ | 21 |
| | | Third Moment, $m_3$ | 22 |
| | | Fourth Moment, $m_4$ | 22 |
| | | Higher Odd Moments | 24 |
| | | Higher Even Moments | 24 |
| | | Measurement of Peak Moments | 24 |
| | | Practical Disadvantages and Uses | 25 |
| | 8 | Manual Peak Area Measurement | 25 |
| | | Height $\times$ Width at Half Height | 25 |
| | | The Condal-Bosch Area | 26 |
| | | Peak Area by Triangulation | 27 |
| | | Manual Measurement of Asymmetric (EMG) Peaks | 28 |
| | 9 | Errors in Peak Area Measurement | 29 |
| | 10 | Accurate Representation of the Solute Profile | 31 |
| | | Peak Distortion | 32 |
| | | Mass and Flow Sensitive Detectors | 34 |
| | | Peak Area and Solute Quantity | 35 |
| | | Linearity | 38 |
| | 11 | Sources of Peak Measurement Error | 41 |
| | | Noise | 41 |
| | | Errors Created by Baseline Drift | 45 |
| | | Errors of Incomplete Peak Resolution | 49 |

|  | Errors from Peak Asymmetry | 59 |
|  | Peak Area vs Peak Height | 63 |
| 12 | References | 66 |

**Chapter 2  Manual Measurement of Peaks**                                **69**

| 1 | Representation of the Detector Signal by Chart Recorder | 69 |
| 2 | Measurement Strategies | 73 |
| 3 | Measurements Based on a Peak Model | 74 |
| 4 | Errors of Manual Measurement | 82 |
|  | Optimum Peak Shape | 84 |
|  | Advantages and Disadvantages of Manual Peak Measurement | 85 |
| 5 | References | 86 |

**Chapter 3  Digital Integrators and Peak Measurement**              **87**

| 1 | A Brief History of Integrators | 87 |
|  | Strip Chart Recorder Techniques | 87 |
|  | Electromechanical Counters | 87 |
|  | Electronic Integrators | 90 |
|  | The Impact of the Microcomputer | 92 |
| 2 | Current Integrator Status | 93 |
|  | The Standard Integrator Specification | 93 |
|  | Integrator Files | 94 |
|  | Analysis Parameters for Peak Measurement | 94 |
|  | Parameters which Override the Integrator's Logic | 97 |
|  | Calculations, Calibrations and Solute Identity | 101 |
|  | Formatting the Analysis Report | 105 |
|  | Report Distribution | 106 |
| 3 | Digital Measurement of Peak Areas | 107 |
|  | Analog to Digital Conversion | 107 |
|  | Data Sampling Frequency | 110 |
|  | Data Bunching | 112 |
|  | Peak Width Parameter and Analysis Reprocessing | 115 |
|  | Peak Sampling Synchronization | 115 |
|  | Rounding or Truncation Errors | 116 |
| 4 | Filtering and Smoothing the Chromatographic Signal | 116 |
|  | Electronic Filters | 117 |
|  | Smoothing | 117 |
|  | Chromatogram Plotting | 123 |
| 5 | Location and Measurement of Peaks | 125 |
|  | Finding the Peaks | 125 |
|  | Properties of the Smoothed Data | 127 |
|  | Location of the Limits of Integration | 129 |
|  | Location of Peak End | 131 |

Measurement of Peak Area 132
Measurement of Peak Height 134
Measurement of Retention Time 134
Measurement of Two Unresolved Peaks 135
Measurement of Fused Groups 137
Shoulders 138
Measurement of Tangent Peaks 138
6 Baselines, A More Detailed Discussion 142
Baseline Drift Limit 142
Formulating a Baseline Definition 144
Mis-timing 'Start' and 'End' 145
End of 'Integrate Inhibit' 146
Forcing Baseline 146
Incorrect Programming of Baseline Drift Tolerance 147
Fused Tangent Measurement 148
Single Peaks on a Rising Baseline 148
Valleys between Fully Resolved Peaks 149
Negative Dips and Constructed Baselines 149
Assigning Baseline beneath the Whole
    Chromatogram 151
7 Conclusions 155
8 References 155

**Subject Index** **157**

# *Glossary of Terms*

| | |
|---|---|
| $A, A_i$ | Area, of component $i$ |
| $B$ | Number of samples in data bunch |
| $B/A$ | Asymmetry ratio |
| $\pm b$ | Baseline gradient |
| $b(n)$ | Coefficient used to evaluate EMG function |
| $c, c_i$ | Concentration, of component $i$ |
| $c_i$ | Weighted coefficient in Savitsky–Golay smoothing |
| $C$ | Capacitance |
| CMR | Common Mode Rejection |
| $E$ | Excess, 4th statistical peak moment |
| $F$ | A/D frequency |
| $f$ | Bunched sample frequency |
| $H$ | Peak height |
| $h$ | Fractional peak height |
| $h_1$ | Height at point of inflection |
| $h_t$ | Detector signal at time $t$ |
| $h'_t$ | 1st derivative of detector signal at time $t$ |
| $h''_t$ | 2nd derivative |
| $I_n$ | Latest measured integral-bunched data sample |
| $I_{(n-M)}$ | '$M$'th predecessor of latest measured integral |
| $I(z)$ | Integral term in EMG function |
| $K$ | Instrument constant |
| $K$ | Solute concentration/detector response |
| $k$ | Area correction factor $= A(\text{true})/A(\text{measured})$ |
| $m$ | $(2m+1)$ data points in a data smoothing window |
| $m_n$ | '$n$'th statistical peak moment |
| $N$ | Noise amplitude |
| $N$ | Savitsky–Golay normalizing constant |
| $N$ | Number of theoretical column plates |
| $n$ | Number of clock pulses equal to one datum |
| $P$ | Number of positive integrals to trigger peak detection |
| $Q$ | Quantity of solute |

| | |
|---|---|
| $R, R_i$ | Response factor, of component $i$ = quantity of solute to produce unit area count |
| $R_s$ | Resolution |
| $R$ | Electrical resistance |
| $S$ | Slope sensitivity |
| $S$ | Normalization scaling factor |
| $S/N$ | Signal to noise ratio |
| $S$ | Total analyte signal |
| $s_i$ | Analyte signal |
| $t$ | General time variable |
| $t_G$ | Retention time of Gaussian peak |
| $t_R$ | Chromatographic retention time |
| $\Delta t$ | Time width of bunched data sample |
| $t_{mean}$ | Mean peak retention time (of centre of gravity) |
| $V$ | Volume |
| $V$ | Valley height |
| $V$ | Voltage |
| $\dot{V}$ | Volume flow rate of mobile phase = $dV/dt$ |
| $W_i, W_s$ | Weights, of solute i, standard |
| $w$ | Peak width |
| $w_b$ | Peak base width |
| $w_h$ | Peak width at fractional height $h$ |
| $w_t$ | Peak width at time $t$ |
| $w_{0.5}$ | Peak width at fractional height $0.5H$ $(= w_h)$ |
| $\pm X$ | Gaussian width boundary |
| $X$ | Time interval in ASTM E 685 noise measurement |
| $y$ | General variable |
| $\bar{y}$ | Smoothed central datum |
| $Z$ | Long term noise amplitude in ASTM E 685 |
| $Z$ | Number of zero slope integrals to end peak measurement |
| $\gamma$ | Skew, 3rd statistical peak moment |
| $\delta$ | Peak separation (dimensionless units) |
| $\sigma, \sigma_G$ | Standard deviation of peak, of Gaussian peak |
| $\sigma/\tau$ | Measure of peak asymmetry from EMG theory |
| $\sigma^2$ | Variance, the 2nd statistical moment |
| $\tau$ | EMG time constant |

CHAPTER 1

# Theory of Peak Measurement

## 1 The Basic Measurements

The basic measurements made by an analyst on a chromatogram for purposes of quantitation are shown in Figure 1.1.

All of these quantities can be obtained from integrator measurements. It is the aim of this book to persuade analysts to use integrators for more than the mere measurement of solute quantities.

The various measurements serve three analytical purposes:

(1) measurement of solute quantity and identification;
(2) experimental validation;
(3) results validation;

and increasingly, for corporate accounting purposes: performance measurement.

### Peak Area and Peak Height

In a controlled analysis, peak area is the true measure of solute quantity always provided that the solute elutes intact and is detected linearly. It can be shown to be so theoretically for peaks with or without a known shape, and experimentally by plots of area against solute quantity.

Peak height is the alternative measure. There is no theoretical proof of this unless a peak shape such as Gaussian is assumed, but experimentally it is easy to show that plots based on height and quantity are also linear over a usable range (Figure 1.2), even for peaks which are not particularly symmetrical as long as their shapes do not change.

The choice of which is best to use in practice is discussed at the end of this section. Area is the normal choice largely because height is susceptible to peak asymmetry while area is not, and the linear dynamic range for area measurement is greater than for height.

1

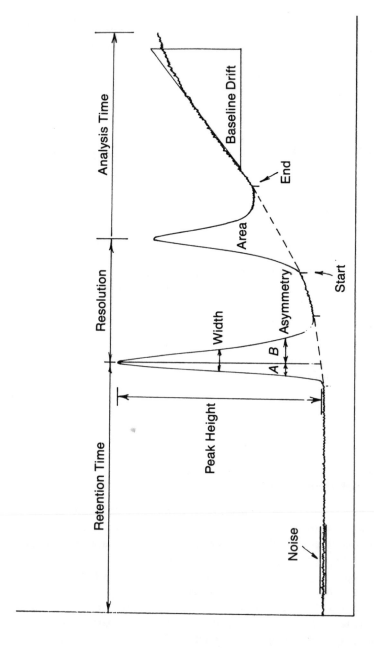

**Figure 1.1**   *The basic measurements*

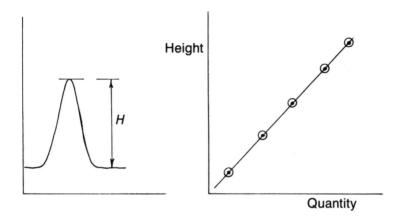

**Figure 1.2**   *Peak height and quantity show a linear relationship*

## Retention Time and Solute Identity

Once solutes in a mixture have been identified, they are subsequently recognized on a daily basis by their retention times. Integrators and computers assign peak names and response factors to peaks which elute in a specified time window. If another peak elutes at that time it will be recognized as the expected peak; if two peaks co-elute they will not be uniquely identified. When retention time varies with sample size due to increasing peak asymmetry, it is quite possible for incorrect identification to be made if a peak crosses from one window into the next. When this happens, the integrator will assign the wrong name, response factor, and standard concentration to that peak.

Integrators can measure relative retention time, *i.e.* the peak retention time compared to the retention time of a standard peak. Experimental variations cancel, and relative retention times are more accurate than absolute retention time but there is the usual problem of finding a suitable standard.

To an integrator, retention time is the elapsed time from the moment of injection until the peak maximum emerges, which includes the gas or solvent hold-up time. The retention time of asymmetric peaks does not coincide with the centre of gravity of the peak. Separation of the observed retention time $(t_R)$ (mode) from the peak centre of gravity (mean) is one measure of asymmetry[1] (see 1st moment and Equation 40).

The mean retention has not achieved any common use in the analytical laboratory. It is difficult to measure manually but integrators which sample the peak signal at a fixed frequency could, with a simple addition to their software, measure it very easily. Its theoretical value lies in the separation of $t_R$ and $t_{mean}$ being equal to the Exponentially Modified Gaussian (EMG) time constant $\tau$ (Figure 1.3).

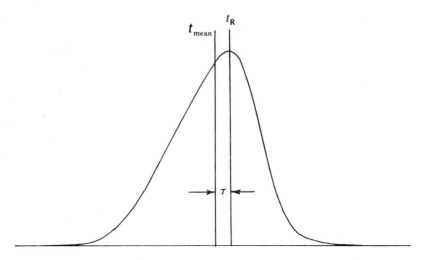

**Figure 1.3**   *Relationship of mean retention time ($t_{mean}$) and observed (mode) retention time ($t_R$) to EMG time constant ($\tau$)*

## 2   Experimental Validation

Most of the instrument checks that an analyst makes before injection to assess readiness can be made automatically by standard integrator routines, or by a BASIC program running in the integrator.

### (a)  Baseline Noise

Absence of noise and a stable baseline are accepted indications that a chromatograph is ready for sample injection. Integrators assess baseline noise by means of 'Noise' or 'Slope Sensitivity' tests, and report its average value. On a daily basis this value should remain reasonably constant for a given analysis. Any deterioration indicates that the column needs reconditioning, the detector needs cleaning, or that some part of the electronics may be about to fail.

### (b)  Baseline Signal Level

Unexpected high signal levels warn the analyst that a peak which ought not to be there is eluting, that the detector has become contaminated, or that temperature or flow control is drifting. Whatever the reason, some delay or corrective action may be necessary before the next injection. Integrators indicate the baseline level and some incorporate a warning if the level is out of range.

### (c)  Peak Width and Column Efficiency

A column is efficient if lateral diffusion of solute bands is restricted during their column residence time. Efficiency is therefore measured in terms of peak width and retention time.

In repeated analyses, the width of a specific peak should be constant.

Variations in width indicate that column performance is varying or some controlling parameter such as temperature is drifting.

Peak width contains components of spreading added by the injection port, system dead volume, and by the detector and electronics as well as the column. If $\sigma^2$ is the peak variance:[1,2]

$$\sigma^2_{total} = \sigma^2_{inj} + \sigma^2_{col} + \sigma^2_{det} + \sigma^2_{etc.} \qquad (1)$$

If the injection technique is not repeatable, or the column is deteriorating, there will be a variation in peak width and apparent variation in column efficiency.

Most integrators can measure peak area and height simultaneously; the ratio of these two quantities gives the peak width at 45.6% of the height of a symmetrical peak $(= \sqrt{(2\pi)}.\sigma)$. Alternatively peak base width is measured as the time interval between the start and end of peak integration $(= 6\sigma$ approximately).

When peak shape is not symmetrical, width can still be used as an empirical measure of column efficiency.

For a Gaussian peak:[3]

$$N = \frac{t_R{}^2}{\sigma^2} \qquad (2)$$

and,

$$\sigma = \frac{w_h}{2\sqrt{[-2\ln{(h)}]}} \qquad \text{from (15) below}$$

thus

$$N = \frac{8t_R{}^2\ln{(h)}}{w_h{}^2} \qquad (3)$$

where
$N$ = number of column plates
$t_R$ = retention time
$w_h$ = width of peak at fractional peak height $h$

and all are quantities the integrator can measure.

There are obvious limitations in measuring column efficiency this way for real peaks, but monitoring the width of a well resolved peak in the chromatogram over successive analyses will highlight imprecise control of instrument parameters, though it will not necessarily identify what is to be improved.

### (d) Peak Asymmetry

Asymmetry makes peaks harder to measure and an aim of method development is therefore to produce symmetrical peaks.

A peak will be symmetrical if:

(i) injection technique is good
(ii) dead volume in the solute flow path is absent

(iii) the solute quantity is not large enough to overload the column or detector

(iv) the residence time of the solute in the stationary phase is long enough to achieve dynamic equilibrium

(v) the solute adsorption isotherm is linear

It follows that where peak asymmetry is evident, some improvements to the analysis may be desirable: change to a different column, reduce the injection volume, *etc.*

Another measure of asymmetry compares the peak half widths on either side of the peak mode (Figure 1.4). This is called the Asymmetry Ratio.[1] Such measurements can in principle be made at any height but are usually made near the peak base where asymmetry is greatest, typically at 10% of peak height for which much theory has been developed,[4] or at 5% for US Pharmacopoeia and FDA requirements. When peaks overlap, the widths at 10% peak height remain measurable for longer, but obtaining the '5% width' demands a better standard of chromatography.

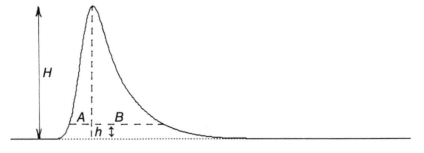

**Figure 1.4**   *Asymmetry ratio*[1] *= B/A. h = 0.05 H or 0.10 H*

Near the top of the peak, width measurements are relatively small, more sensitive to measurement errors, and $B/A$ is closer to unity. Although they may be less precise, it has been suggested[5] they are probably more accurate.

Measurement of the Asymmetry Ratio is now a standard routine on some integrators. It can be applied to any peak in the chromatogram, but it should only be applied to fully resolved peaks.

The asymmetry measurement, $B/A$, as shown in Figure 1.4 is used in the evaluation of mean retention time, the number of column plates and variance derived from the Exponentially Modified Gaussian peak model.[4]

**(e) Peak Resolution**

Peak resolution is the degree of separation of two adjacent peaks. It compares the actual peak top separation to the average constructed peak base width and is used to judge whether the separation between two peaks is good enough for accurate quantitation and fast analysis. In Figure 1.5:

$$\text{Resolution, } R_s = \frac{t_{R2} - t_{R1}}{0.5\,(w_2 + w_1)} \qquad (4)$$

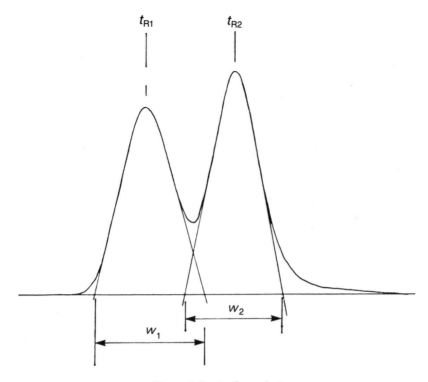

**Figure 1.5** *Peak resolution*

Allowing that the constructed base width of a symmetrical peak is approximately twice the width at half height $w_{0.5}$:

$$R_s \approx \frac{t_{R2} - t_{R1}}{w_{0.5,1} + w_{0.5,2}} \tag{5}$$

In theory, a resolution of 1 means the peaks are just resolved down to baseline and areas can be measured accurately. In practice, if peaks are unequal in size and asymmetric, a resolution nearer 2 is needed to obtain near baseline separation, and even that is sometimes not enough.

Equation 5 is only an approximation and can be as much as 100% in error for asymmetric peaks, but resolution of peaks calculated from the difference in their retention times and widths (derived from Area/Height) can be monitored as an empirical test to show when the column needs reconditioning or replacement.

### (f) Retention Time Stability

The stability of retention time values from one analysis to the next shows up the precision of flow or temperature control or solvent composition. For validation, a limit may be imposed on the amount of variation that is allowed.

# 3   Results Validation

No analysis result should be accepted unquestioningly. Before accepting any results from an integrator or computer, certain simple checks should be made to test the credibility and accuracy of the report.

**(a)  Correct Peak Measurement**
An integrator might draw a baseline in the wrong place or separate two peaks by skimming a tangent where a perpendicular would be better, or *vice versa*. Measurement diagnostics are printed beside peak areas on the integrator's final report describing how the peak was measured. They should be checked and judged correct or not. Event marks showing the start and end of integration should be printed on every wanted peak in the chromatogram.

**(b)  Total Peak Area**
The total peak area count is a measure of injection volume. Its constancy over a sequence of analyses is a measure of injection proficiency and indicates a good or bad injection.

**(c)  Correct Number of Peaks**
In QC analyses, the correct number of peaks occurring at the correct retention times verifies that the correct sample has been injected. After noise peaks and other debris have been removed by the Minimum Peak Size filter, the remaining peak count should be what is expected.

**(d)  Comparison against Standards**
Results may be compared against a standard specification in order to reject or accept a sample if all other aspects of the analysis have been found satisfactory.

# 4   Performance Measurement and Accountability

Accountancy checks on laboratory productivity and instrument utilization are not new but the recent spread of LIM systems is making them more commonplace and thorough. Integrators can be linked to corporate mainframes to provide corporate managers with information for performance measurement. Analysts should be aware of this if only to ensure that the information is not interpreted to their disadvantage.

Where an analysis is an essential part of product quality control, FDA guidelines require the analyst's name to be part of the Final Report.[6] GLP and NAMAS have a similar requirement.

**(a)  Chromatograph Utilization**
The sum of the analysis times gives the daily utilization of a chromatograph. It may be expressed as a fraction or percentage of the working day. Spare

capacity can be noted and instruments used more efficiently. When little spare capacity remains it may be time to invest in another chromatograph.

**(b) Cost per Analysis**
The total number of analyses carried out in a set period divided by the cost of running the laboratory during that period is the cost per analysis.

**(c) Analyst Work Load**
When analysts are assigned to work with specific instruments the number of analyses made each day can be counted and compared with other analysts or against historic work performance.

The cost/analysis and workload ought to reflect the complexity of analyses which might include extraction, derivatization, *etc*. They are estimates which can be used for comparison with other, similar laboratories.

Tests like these are increasingly used in laboratories as legislation and corporate economics call for greater analytical accountability. Some of the laboratory reports generated in future will not be routinely available to the analysts involved.

# 5 Chromatographic Peaks and the Gaussian Function

If a solute is injected as a very narrow band in to a column and partitions linearly between the stationary and mobile phases, the solute distribution inside the column can be represented by a normal or Gaussian distribution curve.[7]

The Gaussian function applies to continuous rather than discrete data[8] and is therefore suited to zones of streaming molecules.

## The Function

When applied to chromatographic peaks it takes the form:

$$h_t = \frac{A}{\sigma.\sqrt{(2\pi)}}.\exp\left[\frac{-(t-t_R)^2}{2\sigma^2}\right] \tag{6}$$

where
$h_t$ = height of peak at time $t$
$A$ = peak area
$t_R$ = time of peak max; the retention time
$\sigma$ = standard deviation of the peak

The curve is bell shaped and symmetrical about the retention time, $t_R$ (Figure 1.6).

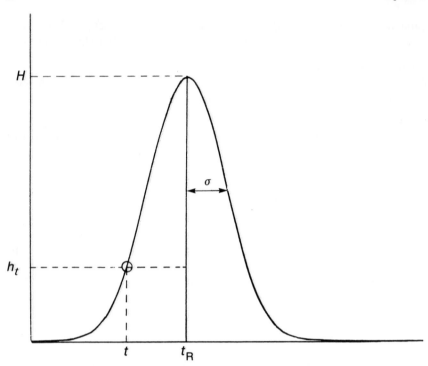

**Figure 1.6**  *Normal or Gaussian distribution curve*

## Peak Height

The first derivative of Equation 6 is given by:

$$h'_t = \frac{-(t - t_R)}{\sigma} \cdot \frac{A}{\sigma\sqrt{(2\pi)}} \cdot \exp\left[\frac{-(t - t_R)^2}{2\sigma^2}\right] \tag{7}$$

$$= \frac{-(t - t_R)}{\sigma} \cdot h_t \tag{8}$$

$h'_t$ is zero at $(t - t_R) = 0$, when $t = t_R$.
At $t = t_R$, the peak is at its maximum height $H$:

$$H = h_{tR} = \frac{A}{\sigma\sqrt{(2\pi)}} \cdot \exp\left[\frac{-(t_R - t_R)^2}{2\sigma^2}\right]$$

$$\therefore \qquad H = \frac{A}{\sigma\sqrt{(2\pi)}} \tag{9}$$

$$= 0.3989\frac{A}{\sigma} \tag{10}$$

Equation 10 is the proof that height is a legitimate substitute for area and a measure of solute quantity for Gaussian peaks.

## Calculation of Fractional Peak Height
Other heights can be expressed in terms of $A$ and $\sigma$ by using equation (10). For example, half height:

$$h = 0.5H = 0.5 \times 0.3989 \frac{A}{\sigma}$$

(11)

$$= 0.1995 \frac{A}{\sigma}$$

and so on.

## Ratio of Area/Height
Some integrators can compute the ratio of measured peak area to peak height.

From Equation 9:

$$\frac{A}{H} = \sigma\sqrt{(2\pi)} = 2.5069\,\sigma \tag{12}$$

which is the peak width at 45.6% of peak height. For a single Gaussian peak $(A/H)^2$ is proportional to peak variance. It will vary with those factors which affect peak width and shape, which is nearly everything, but unless it can be maintained constant, height cannot be used as a substitute for area.

## Calculation of Peak Width at Various Heights
Combining Equations 6 and 9 gives:

$$h_t = H.\exp\left[\frac{-(t - t_R)^2}{2\,\sigma^2}\right] \tag{13}$$

in which the quantity $(t - t_R)$ is equal to half the width of the peak at $t$.

$$\text{i.e. } (t - t_R) = w_t/2 \tag{14}$$

Substituting this into Equation 13 and rearranging gives:

$$w_t = 2\sigma \sqrt{\left[-2\ln\frac{h_t}{H}\right]} \tag{15}$$

Figure 1.7 summarizes the heights and widths at key peak locations on a Gaussian Peak.

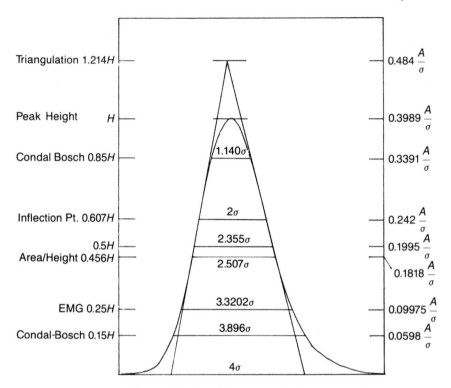

**Figure 1.7**   *Key dimensions of a Gaussian peak*

### Gaussian Peak Shape Tests

(i)  The Asymmetry Ratio, $B/A = 1$ at all heights, most importantly near the base.

(ii) The ratios of peak widths at various heights will have known values; *e.g.* at 50%, 30%, and 10% the peak widths will have the ratios 0.5487:0.7231:1.

### The Points of Inflection
The points of inflection are the points of maximum slope on each side of the peak. It is through these points that tangents are drawn for the 'triangulation' method of peak area measurement.

The points of inflection are located where the second derivative $h''_t = 0$.

$$h''_t = \left[ \frac{(t - t_R)^2}{\sigma^2} - 1 \right] . \frac{A}{\sigma\sqrt{(2\pi)}} . \exp \left[ \frac{-(t - t_R)^2}{2\sigma^2} \right] \tag{16}$$

$$= \left[ \frac{(t - t_R)^2}{\sigma^2} - 1 \right] . h_t \tag{17}$$

and this is zero when:

$$\frac{(t - t_R)^2}{\sigma^2} - 1 = 0$$

*i.e.* when

$$t - t_R = \pm \sigma \tag{18}$$

Between the points of inflection the peak width is $2\sigma$.

## Height of the Peak at the Point of Inflection

This is found by substituting $(t - t_R) = \sigma$ in Equation 6:

$$h_I = \frac{A}{\sigma\sqrt{(2\pi)}} \cdot \exp\left[\frac{-\sigma^2}{2\sigma^2}\right]$$

$$= 0.242 \frac{A}{\sigma} \tag{19}$$

from Equation 10, $\qquad\qquad = 0.6067 \, H$

## Maximum Peak Slope

The maximum peak slope is determined by making the same substitution, $t - t_R = \sigma = 1$, into Equation 7 for the first derivative, $h'_I$:

$$h'_I = \frac{\pm \sigma}{\sigma} \cdot \frac{A}{\sigma\sqrt{(2\pi)}} \cdot \exp\left[\frac{-\sigma^2}{2\sigma^2}\right]$$

$$= \pm 0.242 \, A \tag{20}$$

At the points of inflection, the height and the gradient have the same numerical value.

## Fractional Peak Area Bounded by Various Widths

The boundaries of a Gaussian peak are theoretically infinite and it is important to know how much area is lost by adopting finite dimensions compatible with observed peak widths.

The quantity of peak area between two boundaries '$-X$' and '$+X$' can be obtained by integrating Equation 6 between these limits, but it is easier to consult statistical tables[9] which give the fractional peak area from $-4\sigma$ to $+4\sigma$.

A summary of the commonly quoted values is given in Table 1.1 and shown in Figure 1.8.

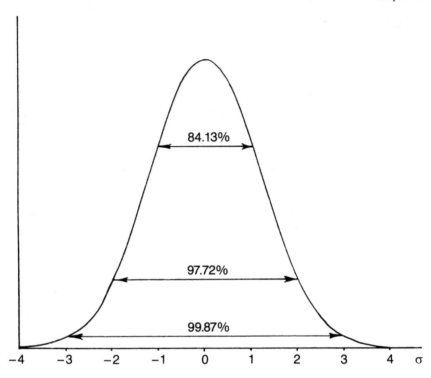

**Figure 1.8**   *Area of Gaussian peak bounded by different widths*

**Table 1.1**

| σ | Area% | σ | Area% |
|---|---|---|---|
| 0.5 | 69.15 | 2.5 | 99.38 |
| 1.0 | 84.13 | 3.0 | 99.87 |
| 1.5 | 93.32 | 3.5 | 99.98 |
| 2.0 | 97.72 | 4.0 | 100.0 |

**Loss of Area from the Base of a Gaussian Peak**

The base of a peak will not be measured accurately if:

(1) the baseline is noisy making peak detection late, after the peak has risen above the noise;
(2) the baseline signal has drifted off the bottom of the chart recorder and cannot be followed by the pen;
(3) integrator slope sensitivity parameter is set too large (too insensitive) so that slope is detected late and lost early;
(4) the baseline has drifted below the lower limit of the integrator's operating range, and only that part of the peak rising above it is measured.

Area lost from the base of a peak is lost from the widest part where it matters most.

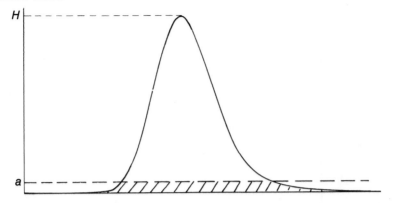

**Figure 1.9** *The shaded area is lost from widest part where it matters most*

Littlewood[10] calculated the percentage area loss from the base of a Gaussian peak using Equation 21:

$$\text{area lost} = 200\left[\text{erfc}\left(2\ln\frac{H}{a}\right)^{\frac{1}{2}} + \frac{a}{H}\left(\frac{1}{\pi}\ln\frac{H}{a}\right)^{\frac{1}{2}}\right] \tag{21}$$

**Table 1.2**

| $a/H\%$ | Area loss % |
|---------|-------------|
| 0.5 | 1.42 |
| 1.0 | 2.66 |
| 2.0 | 4.98 |
| 3.0 | 7.10 |
| 5.0 | 11.20 |

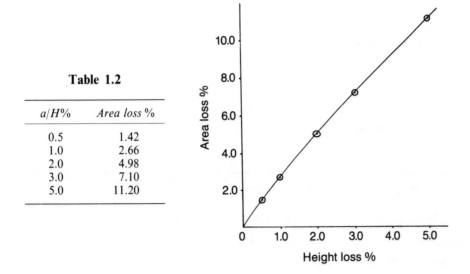

**Figure 1.10** *Effect of losing peak base. More area is lost than height*

**Gaussian Peak Maximum as an Approximate Parabola**

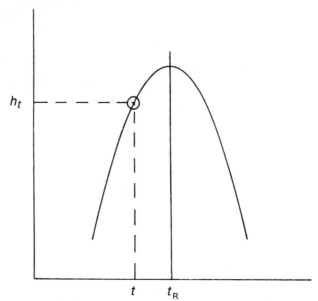

**Figure 1.11**   *Gaussian peak top is parabolic*

For a Gaussian shaped peak:

$$h_t = H.\exp\left[\frac{-(t - t_R)^2}{2\sigma^2}\right] \qquad \text{from (13)}$$

An exponential function $\exp(x)$ can be written as the series:

$$\exp(x) = 1 + x + \frac{x^2}{2!} + \frac{x^3}{3!} + \cdots + \frac{x^n}{n!} \qquad (22)$$

and by comparison with Equation 13

$$x = \frac{-(t - t_R)^2}{2\sigma^2} \qquad (23)$$

so Equation 22 may be re-written

$$\exp\left[\frac{-(t - t_R)^2}{2\sigma^2}\right] = 1 - \frac{(t - t_R)^2}{2\sigma^2} + \frac{(t - t_R)^4}{8\sigma^4} - \cdots + \frac{(t - t_R)^{2n}}{(2\sigma)^{2n}.n!} \qquad (24)$$

Close to the peak maximum, $(t - t_R) < \sigma$ and Equation 24 rapidly converges allowing terms of the 4th order and higher to be neglected. Approximately:

$$\exp\left[\frac{-(t - t_R)^2}{2\sigma^2}\right] = 1 - \frac{(t - t_R)^2}{2\sigma^2} \tag{25}$$

Substituting Equation 25 into Equation 13 gives:

$$h_t = H\left[1 - \frac{(t - t_R)^2}{2\sigma^2}\right] \tag{26}$$

and this has the form of a parabola (Figure 1.11).

When a more accurate representation over a wider range is required, the peak maximum is fitted to a polynomial (see Chapter 3). This allows more data samples to be included, and is how integrators measure retention time.

# 6 The Exponentially Modified Gaussian Function[11]

Not many chromatographic peaks are Gaussian and this has led to much effort since 1959 to find a better peak model.[11-17]

The exponentially modified Gaussian, or EMG function, gives good agreement between theory and experiment in many real cases.[65] It is a Gaussian function convoluted (bent) on to an exponential axis of time constant τ.

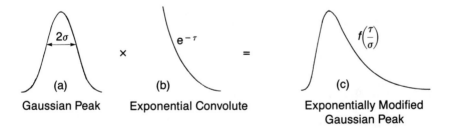

| (a) | (b) | (c) |
| Gaussian Peak | Exponential Convolute | Exponentially Modified Gaussian Peak |

**Figure 1.12** *The exponentially modified Gaussian function*

The equation for the EMG function can be expressed in several ways; a typical example is:

$$h_t = \frac{A}{\tau}\exp\left[\frac{1}{2}\left(\frac{\sigma_G}{\tau}\right)^2 - \left(\frac{t - t_G}{\tau}\right)\right]\int_{-\infty}^{z}\frac{1}{\sqrt{(2\pi)}}\cdot\exp(-y^2/2)\,dy \tag{27}$$

where                          $A$ = area of EMG peak

$t_G$ = retention time of Gaussian peak

$\sigma_G$ = standard deviation of Gaussian peak

$\tau$ = time constant of exponential axis

$$z = \frac{t - t_G}{\sigma_G} - \frac{\sigma_G}{\tau} \tag{28}$$

and                            $y$ = dummy variable of integration

In order to evaluate $h_t$, the integral term in Equation 27 is replaced in one of two ways:

$$\text{if } I(z) = \int_{-\infty}^{z} \frac{1}{\sqrt{(2\pi)}} \cdot \exp\left(\frac{-y^2}{2}\right) dy$$

(a)   $I(z)$ can be replaced by an error function:

$$I(z) = \text{erf}\left(\frac{1}{\sqrt{(2\pi)}} \cdot \left[\frac{t - t_G}{\sigma_G} - \frac{\sigma_G}{\tau}\right]\right) \tag{29}$$

$h_t$ is usually evaluated by computer, and if the error function is or can be made a standard routine of the computer, evaluation of $I(z)$ and Equation 27 is easily carried out.

(b)   $I(z)$ can be replaced by the approximations:

$$I(z < 0) = \frac{1}{\sqrt{(2\pi)}} \cdot \exp\left(\frac{-z^2}{2}\right) \sum_{n=1}^{5} \frac{b(n)}{(1 + pz)^n} \tag{30}$$

and $I(z > 0) = 1 - I(z < 0)$ (31)

in Equation 30, $z$ is defined in Equation 28 and:

$p = 0.2316419;$
$b(1) = 0.319381530;$
$b(2) = -0.356563782;$
$b(3) = 1.781477937;$
$b(4) = -1.821255978;$
$b(5) = 1.330274429$

A computer program to evaluate Equations 30 or 31 and hence $h_t$ is straightforward. One for the Apple II microcomputer has been published.[11]

An alternative equation[13] for the exponentially modified Gaussian function in a form that does not involve an integral term is:

$$\frac{h_t}{H} = \sqrt{\left(\frac{\pi}{2}\right)} \frac{\sigma_G}{\tau} \exp\left[\frac{-\sigma_G}{2\tau}\left(\frac{t - t_R}{\sigma_G} - \frac{\sigma_G}{\tau}\right)\right] \cdot \text{erf}\left[\frac{1}{\sqrt{2}}\left(\frac{t - t_R}{\sigma_G} - \frac{\sigma_G}{\tau}\right)\right] \quad (27a)$$

## Practical Application of the EMG Function

Adoption of the EMG function in analytical laboratories has been held at bay by the complexity of Equation 27. However, useful (and simple) formulae for chromatographic quantities have been derived[4,18] for asymmetric peaks where $\tau/\sigma < 3$, some involving the asymmetry ratio $B/A$: These formulae include four alternative area equations:

Peak Area $A = 0.753\, H\, w_{0.25}$ (32)

or, $\qquad = 0.586\, H\, w_{0.1}(B/A)^{-0.133}$ (32a)

or, $\qquad = 1.07\, H\, w_{0.5}(B/A)^{0.235}$ (32b)

or, $\qquad = 1.64\, H\, w_{0.75}(B/A)^{0.717}$ (32c)

Theoretical Plates $N = \dfrac{t_R^2}{\sigma_G^2 + \tau^2} = \dfrac{41.7\left(\dfrac{t_R}{w_{0.1}}\right)^2}{B/A + 1.25}$ (33)

Variance $\sigma^2 = \dfrac{t_R^2}{N} = \sigma_G^2 + \tau^2$ (34)

where $\qquad H$ = peak height
$w_{0.25}$ = peak width at 25% of height
$w_{0.75} =$ „ „ „ 75% „ „
$w_{0.5} =$ „ „ „ 50% „ „
$w_{0.1} =$ „ „ „ 10% „ „

and $B/A$ is the asymmetry ratio at 10% height in $\tau/\sigma$ range 1.1 to 2.8.

Where peaks fit the EMG function, Equation 32 is very convenient for manual measurement of area. It does not depend on $B/A$ and is as easy as height $\times w_{0.5}$ to calculate.

## EMG Peak Shape Tests

(i) Peaks should not be symmetrical: the asymmetry ratio, $B/A <> 1$. A better test is $B/A > 1.1$.[4]

(ii) Peak areas calculated from Equations 32a, b, and c should agree in theory within about 2% of the area calculated from Equation 32. Larger errors might perhaps be compared with the results from a Gaussian shape test before abandoning Equation 32 and reverting to Gaussian calculations.

# 7 Statistical Moments of a Chromatographic Peak[19,20,21,65]

Instead of defining asymmetric peak shapes by mathematical models such as Exponentially Modified Gaussian, an alternative approach is to apply measurement techniques that assume no peak shape.

Such a technique is the 'method of statistical moments' which characterizes those distributions that cannot be expressed by known curves. Application of statistical moments to chromatography requires introduction of the concepts of finite peak boundaries, peak overlap and non-linear baseline which have no meaning in the original context of a distribution function. The results must also be compared with other models to see how they agree and differ.

There is an infinite number of peak moments but only the first five are used in connection with chromatographic peaks. The general formula for them, is:

$$m_n = \frac{\int_{-\infty}^{+\infty} t^n h_t \, dt}{\int_{-\infty}^{+\infty} h_t \, dt} \tag{35}$$

which is normalized to the zeroeth moment ($n = 0$).

## Zeroeth Moment, $m_0$

This is the peak area. In Equation 35, $n = 0$:

$$m_0 = \frac{\int_{-\infty}^{+\infty} t^0 . h_t \, dt}{\int_{-\infty}^{+\infty} h_t \, dt} = 1 \text{ (since } t^0 = 1) \tag{36}$$

Equation 36 reflects the common practice of statisticians to normalize population totals to 1 or 100%. In chromatography the non-normalized moment is more useful:

$$m_0 = \int_{-\infty}^{+\infty} t^0 . h_t \, dt = \text{Peak Area, } A \tag{37}$$

## First Moment, $m_1$

The first moment is the 'mean retention time', or retention time measured at the centre of gravity of the peak. It is different from the chromatographic retention time measured at the peak maximum unless the peak is symmetrical (Figure 1.13).

$$m_1 = \frac{\displaystyle\int_{-\infty}^{+\infty} t^1 . h_t \, dt}{\displaystyle\int_{-\infty}^{+\infty} h_t \, dt} \tag{38}$$

in non-normalized form:

$$m_1 = \frac{1}{m_0} \int_{-\infty}^{+\infty} t^1 . h_t \, dt \equiv t_R + \tau \tag{39}$$

where $t_R$ = retention time of a Gaussian peak and $\tau$ = EMG time constant.

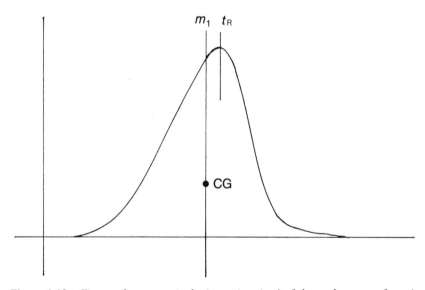

$m_1$ $t_R$

CG

**Figure 1.13** *First peak moment is the 'retention time' of the peak centre of gravity*

In chromatography, $m_1$ can be referenced to the peak retention time, $t_R$, in which case Equation 39 modifies to:

$$m_1 = \frac{1}{m_0} \int_{-\infty}^{+\infty} (t - t_R) h_t \, dt \tag{40}$$

## Second Moment, $m_2$

The second moment is the peak variance, $\sigma^2$, where $\sigma$ is the peak standard deviation.

$$m_2 = \frac{1}{m_0} \int_{-\infty}^{+\infty} t^2 h_t \, dt \qquad \equiv \sigma_G^2 + \tau^2 \tag{41}$$

The variance of a chromatographic peak, $\sigma^2$, is a measure of lateral spreading. It is the sum of the variance contributed by different parts of the instrument system. Generally:

$$\sigma_{\text{Total}}^2 = \sum_i \sigma_i^2 \tag{42}$$

The chromatographic interpretation of which is:

$$\sigma_{\text{Peak}}^2 = \sigma_{\text{inj}}^2 + \sigma_{\text{col}}^2 + \sigma_{\text{det}}^2 + \sigma_{\text{etc.}}^2 \tag{43}$$

## Third Moment, $m_3$

The third moment describes vertical asymmetry, or skew. It is a measure of the departure of the peak shape from the Gaussian standard.

In normalized form:

$$m_3 = \frac{1}{m_0} \int_{-\infty}^{+\infty} t^3 . h_t \, dt \equiv 2\tau^3 \tag{44}$$

Skew is sometimes expressed[1,12,16] as the dimensionless quantity $\gamma$:

$$\text{Skew, } \gamma = \frac{m_3}{m_2^{3/2}} = \frac{2\left(\dfrac{\tau}{\sigma_G}\right)^3}{\left[1 + \left(\dfrac{\tau}{\sigma_G}\right)^2\right]^{3/2}} \tag{45}$$

A symmetrical peak has a skew of zero. Peaks which tail have positive skew and their first moment is greater than the peak retention time (Figure 1.14). Peaks which front have negative skew; their first moment is less than the retention time.

## Fourth Moment, $m_4$

The fourth moment or 'excess', is a measure of the compression or stretching of the peak along a vertical axis, and how this compares to a Gaussian standard for which $m_4 = 0$. It can be visualized by moving in or pulling apart the sides of a Gaussian peak while maintaining constant area.

$$m_4 = \frac{1}{m_0} \int_{-\infty}^{+\infty} t^4 h_t \, dt \tag{46}$$

$$\equiv 3\sigma^4 + 6\sigma^2\tau^2 + 9\tau^4$$

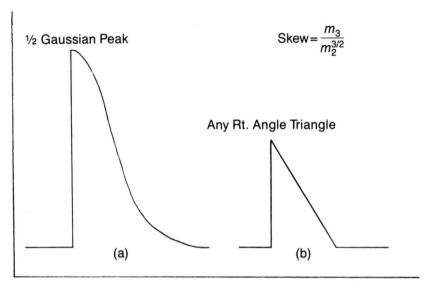

**Figure 1.14** *Examples of skew.* (a) $\gamma = 0.995$; (b) $\gamma = 1.131$
(Using data from *J. Chem. Phys.*, 1963, **38**, 437)

If the peak is compressed or squashed down in comparison, (*i.e.* if the detector overloads) its excess is negative (Figure 1.15). If it is taller, its excess is positive.

Excess $(E)$, too, can be expressed in the dimensionless form:[12]

$$E = \frac{m_4}{m_2{}^2} - 3$$

For Gaussian peaks $m_4/m_2{}^2 = 3$.

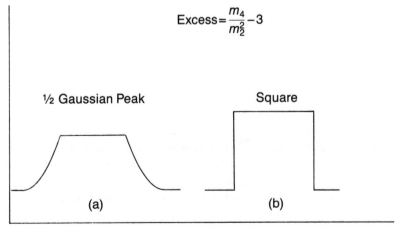

**Figure 1.15** *Examples of excess.* (a) $\frac{1}{2}$ *Gaussian peak*, $E = -1.377$; (b) *square peak*, $E = -1.666$
(Using data from *J. Chem. Phys.*, 1963, **38**, 437)

## Higher Odd Moments

All higher odd moments are measures of vertical asymmetry, like skew, and for symmetrical peaks they are all zero.

## Higher Even Moments

Higher even moments have no similar easy interpretation except that like the second and fourth moments they are measures of lateral dispersion.

## Measurement of Peak Moments

Peak moments lend themselves very well to measurement by computer.

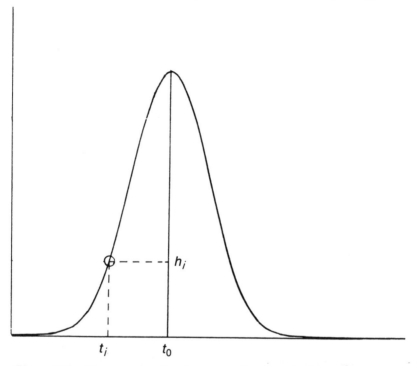

**Figure 1.16** *Measurement of peak moments. Position of reference time* $t_0$ *is arbitrary*

Moments are calculated from measurements of $h_i$ made at regular time intervals over the whole peak, which is exactly what integrators do.

Integrator measurement of peak area is the same as measurement of the zeroeth moment. Modal retention time is measured in preference to the first moment, and skew, if it is measured at all, is measured as the asymmetry ratio, $B/A$. Integrators only measure peak width and not yet excess.

Should these quantities ever be routinely wanted by analysts, integrators could provide them by simple additions to their software.

## Practical Disadvantages and Uses

Moments are notoriously susceptible to noise.[23,24] With the exception of the zeroeth moment, $S/N$ ratio must be better than $100$[25] for the measurements to be meaningful. They are severely affected by baseline drift, inaccurate determination of the peak limits, tailing, incomplete resolution, and insufficient sampling frequency.

If peak moments higher than zero have a future use for the analyst, it would seem to be as diagnostics in high quality chromatography. They are too sensitive for routine use.

The mean retention can be compared with the chromatographic retention to measure asymmetry. Variance can be used to estimate system stability, monitor column suitability, performance and deterioration.

Skew shows column suitability also. If too many peaks are skewed, an alternative column might be a better choice. Excessive or increasing skew is an indicator of column deterioration or of system errors such as column overload or adsorption of solutes in the injector.

Excess would indicate detector overloading or saturation.

# 8 Manual Peak Area Measurement

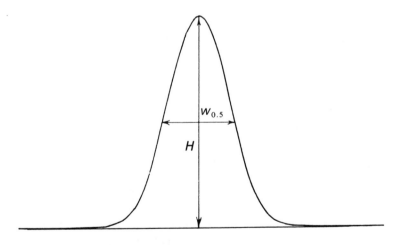

**Figure 1.17** *Manual measurement of a Gaussian peak. Area = $kw_{0.5}H$. For a Gaussian peak $k = 1/0.939$*

## Height × Width at Half Height

The area of a Gaussian peak can be calculated from the product of its height and width at half height, which is obtained by substituting $h_t/H = 0.5$ into Equation 15:

$$\text{Area} = H.w_{0.5} \tag{47}$$

$$w_{0.5} = 2\sigma\sqrt{[-2\ln(0.5)]}$$
$$= 2.3548\,\sigma \tag{48}$$

$$\text{Area} = 0.3989\frac{A}{\sigma} \times 2.3548\,\sigma$$
$$= 0.939\,A \tag{49}$$

*i.e.* the computed area is 93.9% of the theoretically true area, from which $k = 1/0.939$ in Figure 1.17.

## The Condal-Bosch Area[26]

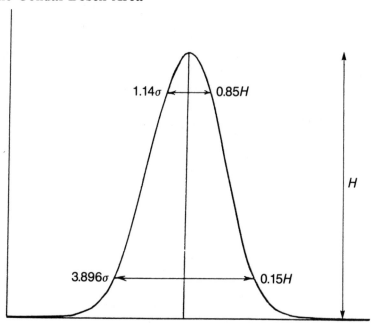

**Figure 1.18** *The Condal-Bosch area*

Instead of the width at half height, Condal-Bosch used the average of the peak widths at 15% and 85% of the peak height.

$$\text{Area} = H.0.5[w_{0.15} + w_{0.85}] \tag{50}$$
$$w_{0.15} = 2\sigma\sqrt{[-2\ln(0.15)]} = 3.896\,\sigma$$
$$w_{0.85} = 2\sigma\sqrt{[-2\ln(0.85)]} = 1.140\,\sigma$$

$$\text{Peak Area} = 0.3989\frac{A}{\sigma} \times \frac{(3.896\,\sigma + 1.140\,\sigma)}{2}$$
$$= 1.004\,A \tag{51}$$

Which is very close to the true peak area.

## Peak Area by Triangulation

The triangle was probably the first peak model to be used. A Gaussian peak can be related to the triangle formed when tangents drawn through the points of inflection intersect the peak baseline and each other.

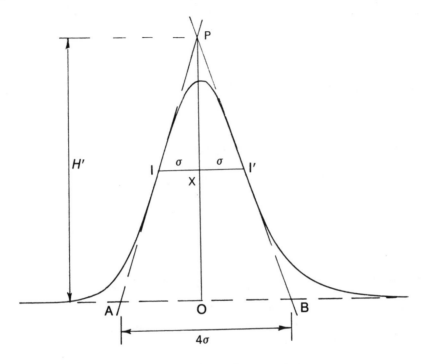

**Figure 1.19**  *Peak area by triangulation*

At the point of inflection, $I$, the gradient of the peak (and therefore of the side of the triangle APB) is $\pm 0.242\,A$. The height of $I$ is $0.242\,A/\sigma$, and the half width of the peak at $I$ is $\sigma$ (see Equations 18, 19 and 20).

$$\text{In Figure 1.19, } OX = 0.242\frac{A}{\sigma} \qquad \text{from (19)}$$

$$\text{gradient} \qquad \frac{PX}{\sigma} = 0.242\frac{A}{\sigma} \qquad \text{from (20)}$$

at $I$, $\sigma = 1$, $PX = OX$,

$$PO = 2OX = 0.484\frac{A}{\sigma} = H'$$

and

$$AO = 2\sigma = OB$$

**Method (1)**
The peak area calculated from the area of the exterior triangle is:

$$\text{Area} = 0.484 \frac{A}{\sigma} \times 2\sigma \tag{52}$$

$$= 0.968\, A \tag{53}$$

or 96.8% of the true peak area.

**Method (2)**
Area from triangular base width and peak height

$$\text{Area} = 0.3989 \frac{A}{\sigma} \times 2\sigma \tag{53a}$$

$$= 0.7978\, A$$

It is implicit in triangulation that the inflection point tangents converge at an angle large enough to allow precise determination of the intersect. If this is not the case it will be easier and more precise to measure the peak height. The measurement of area by Equation 53a should be more precise than by method (1), even though less accurate.

## Manual Measurement of Asymmetric (EMG) Peaks

(1) Foley's equation (see Equation 32) for the area of a resolved exponentially modified Gaussian peak requires no overt measurements of asymmetry such as $B/A$ yet covers a range of asymmetry up to $\tau/\sigma = 3$. It also fits a symmetrical peak:

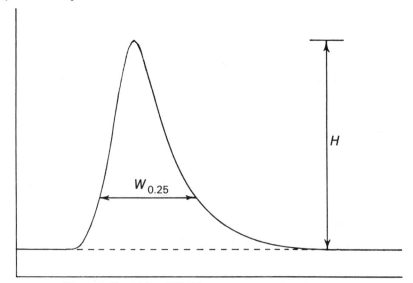

**Figure 1.20**  *Foley (EMG) measurement of peak area*

$$\text{Area} = 0.753.H.w_{0.25} \tag{32}$$

if Gaussian, $$= 0.753 \times 0.3989 \frac{A}{\sigma} \times w_{0.25}$$

From Equation 29, $\quad w_{0.25} = 2\sigma\sqrt{[-2\ln(0.25)]} = 3.3302\,\sigma$
$$\text{Area} = 1.0003\,A \tag{54}$$

which is even closer to the true area than the Condal-Bosch calculation, requires fewer measurements, and suffers less from overlap. It is more accurate for the measure of symmetrical peaks than the more commonly used height × half width.

(2) When peak overlap prevents the accurate measurement of $w_{0.25}$, an alternative measure of EMG peak area can be made from the width at 75% of peak height, $w_{0.75}$, the peak height and the asymmetry ratio, $B/A$ at that height:[4]

$$A = 1.64\,H.w_{0.75}.(B/A)^{0.717} \tag{32c}$$

The accuracy is within 4% for computer simulated overlapping peaks where the valley rises to $0.45\,H$ of the smaller peak. Equation 32c applies to the measurement of either peak for asymmetries up to $\tau/\sigma < 3$.

Applied to symmetrical peaks, $B/A = 1$, $w_{0.75} = 1.517\sigma$, and the measured area is 99.2% of the true area.

The EMG function is largely used for the study of variance, skew, and column plate measure and the errors of these measurements have ranged up to 30%.[27] As a simple manual measure of skewed peak area it has been found to be experimentally accurate to within 10%.[4,28]

# 9 Errors in Peak Area Measurement

Modern chromatographs using stable bonded stationary phases and automatic injection can produce a set of closely similar results from the repeated analysis of a particular sample. The relative standard deviation or coefficient of variation of these results may be very small (and therefore pleasing), while the results themselves are quite different to known answers.

This stability highlights the difference between precision and accuracy.

**Accuracy**
Accuracy is a measure of how close the experimental result is to the 'true result'. The difference between the 'true' and experimental result is called the bias.

**Precision**
Precision is a measure of how close the results from repeated experiments are to each other.

A graphical interpretation of precision and accuracy is given in Figure 1.21:

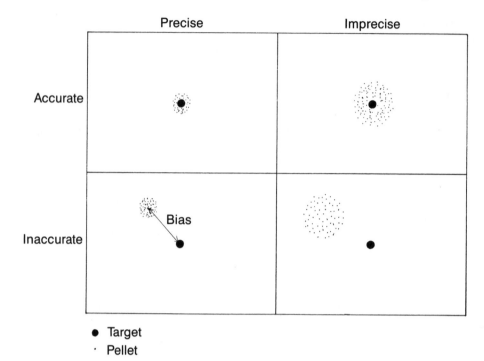

**Figure 1.21**    *Accuracy and precision*

Because of the quality of modern manufacturing techniques, chromato-graphic instrumentation can achieve a precision which analysts may mis-interpret as accuracy. The same result from three consecutive analyses does not mean it is the correct result.

Method development must demonstrate accuracy as well as precision and it will do so by calibration and validation.

Chromatographic analyses encounter the two types of error:

**Random errors**
These are unpredictable or indeterminate in nature and affect the precision of analysis. Coefficients of variation are a measure of random error. Statistical methods of signal improvement are used to remove random error.

**Systematic errors**
Systematic errors affect the accuracy of the result. They can be fixed in size or vary with the amount of sample analysed. Unless the true experimental result is known, or the behaviour of the systematic error is predictable, a measure of this error cannot be obtained from the experiment. Systematic error is often recognized only after an independent result is obtained which differs from the observed result by an amount too great to be explained by random experimental error.

**Repeatability**

Repeatability is the closeness of results between consecutive analyses carried out on the same chromatograph by the same operator working in constant conditions. It is therefore the same as precision.

**Reproducibility**

Reproducibility is the closeness between results from analyses of the same solution by different people working with different chromatographs.

Reproducibility rarely achieves the precision of repeatability. Published data on comparisons of the precision of various forms of peak measurement usually imply repeatability.

**Causes of Imprecision**

Imprecision (repeatability) means that measurements differ from one analysis to the next. Reasons for this are attributed to imperfect sample preparation and injection technique, solute adsorption or decomposition, solvent leaks, columns which need conditioning, instrument instability and, generally, the kind of problems which analysts feel they should be able to solve.

**Causes of Inaccuracy**

There are only two causes of inaccuracy:

(1) the detector signal does not accurately represent the solute profile (it is assumed that the sample represents the original solution);

(2) peaks are measured according to incorrect rules. For example when asymmetric peaks are measured by methods derived from Gaussian theory.

Solute quantities can be measured accurately if:

(a) solutes are eluted intact and are detected. No chemical reactions or irreversible adsorption have taken place en route;

(b) the solute peaks are symmetrical and fully resolved, on a flat and noise free baseline;

(c) analysis control parameters such as mobile phase flow rate and composition and column or detector temperatures are stable and precise;

(d) the detector and integrator are working within their linear operating ranges;

(e) the peaks are measured accurately.

Only two of these factors, (d) and (e), concern the integrator; the accuracy and precision of an analysis are largely determined by what happens before the detector signal reaches the integrator. No analyst should expect an integrator to compensate for poor chromatography.

# 10   Accurate Representation of the Solute Profile

Measured peak area or height is proportional to solute quantity only under certain detector operating conditions. When these conditions are not met the detector does not accurately represent the profile of the eluting solute zone; what the analyst sees and what the integrator measures is not what the column delivers.

The detector signal is the arithmetic sum of the component signals from all detected species passing through the detector cell.

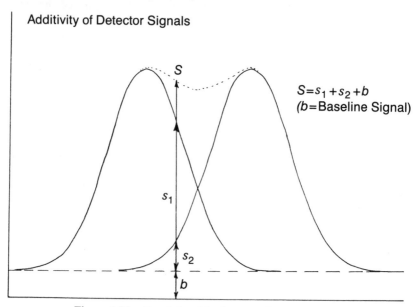

**Figure 1.22** *Total signal is the sum of component signals*

When unresolved groups of solutes pass through the detector, the output signal is the sum of the component signals from each detected solute, and from stationary phase and mobile phase too if these produce a measurable response (Figure 1.22):

$$S = \sum_{o}^{n} r_i c_i \tag{55}$$

where $r_i$ is the response of unit component of the solute and $c_i$ = concentration or amount of solute.

Peaks don't overlap in the sense that one covers or hides another, they sit on top of each other so that the whole signal is visible. This does not mean every peak can be seen—the change from packed columns to WCOTs showed how peaks can be concealed; there is a limit of detection associated with resolution as well as peak size.

The absence of distinguishing features such as peak maxima has resulted in the failure of past efforts to separate overlapping peaks mathematically.

## Peak Distortion

The peak which appears on a strip chart recorder has passed through the recorder's electronics and been drawn by a pen of finite inertia. The peak drawn by an integrator has been digitized, smoothed, reconstructed through a D/A converter and finally plotted. These operations distort the original solute profile.

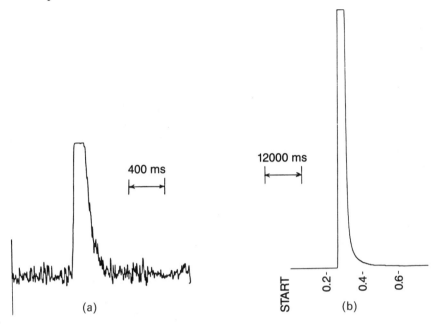

**Figure 1.23** (a) *Oscilloscope trace of FID Peak;* (b) *Integrator plot of same peak*

Figure 1.23a shows an FID peak recorded by an oscilloscope. Figure 1.23b shows the same peak after it has passed through an electrometer and integrator and been filtered for noise by both. The difference in peak width and absence of noise are obvious. Noise is removed by a combination of capacitive filters and software soothing in the detector and the integrator.

The appearance of the gas chromatographic peak was largely established in the 1960s by the electronics of that time. The peaks looked like Figure 1.23b instead of Figure 1.23a because the detectors and electronics represented them that way. Modern electronics are much faster and are able to represent the peak accurately as in Figure 1.23a, but no analyst would buy such a chromatograph, so manufacturers make chromatographs and integrators to filter noise and show peaks as icons.

Distortion by capacitive (RC) filters skews peak shape. Peaks are broadened, heights are reduced and the reduction may be, to some extent, a function of retention time for fast peaks if peak width increases (*i.e.* peak 'frequency' decreases) with retention. For example using an old GC with the latest WCOT columns can result in the elution of early peaks which are narrower than the manufacturer allowed for when the amplifier and detector were designed, with the result that they are filtered or attenuated as noise in a way that is different to later, broader peaks which are more compatible with the time constant.

The redeeming feature is that peak area is not changed by RC filtering and remains a measure of solute quantity. In consequence, when very narrow peaks are being measured, integrators should be primarily regarded as area measuring devices.

UV detectors for LC are less noisy than FIDs. Their time constants are set to approximately 1/20 of peak base width.[29,30] Some have a selection of flow cells for different 'LC speeds' and their time constant can be changed by the user over a limited range.

If height is measured, then peaks which are compared to each other should have similar widths or a calibration on height must be made. When integrators are used without calibration, it is safer to use area.

Strip chart recorders, with fast linear response and small pen head inertia, distort peak shapes less than integrators. If a chromatographer is more interested in peak shape than size, the recorder may be a more accurate instrument of display.

## Mass and Flow Sensitive Detectors[31]

Detectors are either flow sensitive or mass sensitive depending on whether the mere presence of solute inside the detector cell is sufficient to create a response, or whether the solute is consumed in a chemical reaction, and it is the reaction or its products that create the response. Solutes pass unchanged through flow sensitive detectors which can therefore be linked in series.

The difference between the two detectors is highlighted by considering the effect of stopping the (single component) mobile phase as a solute passes through the detector cell.

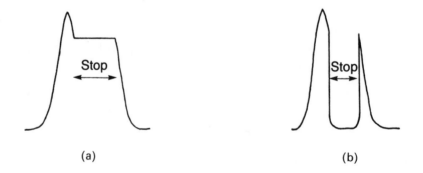

(a)                                                                      (b)

**Figure 1.24**  *Effect of stopping mobile phase on detector signal.* (a) *Flow sensitive;* (b) *Mass sensitive*

In a flow sensitive detector, the detector signal is held constant and it will only change when the flow re-starts and sweeps the solute from the cell. The area of the solute peak is increased but the height remains nearly constant (in fact a 10% reduction in flow has been observed to produce a 1% increase in height[32]).

In contrast, the signal level from a mass sensitive detector falls to zero (at a rate determined by the detector time constant) as the solute inside the cell is consumed and no more is delivered. When flow is continued the signal picks up again as the remaining solute is swept into the detector cell.

The peak area is split, but the sum of the two split areas is the same as the peak area if the flow had not stopped.[33] Peak height is not preserved unless the maximum has eluted before the flow was interrupted.

The two types of detector are summarized in Table 1.3.

**Table 1.3**

| Gas Chromatography | | Liquid Chromatography | |
|---|---|---|---|
| FID | Mass | UV or UV/VIS detector | Flow |
| TCD | Flow | Refractive Index detector | Flow |
| ECD | Flow (Mass) | Fluorimetric detector | Flow |
| N/P FID | Mass | Electrochemical detector | Mass |
| FPD | Mass | Coulometric detector | Mass |
| Mass Spectrometer | Mass | | |
| Hall detector | Flow | | |
| He ID | Mass | | |

When flow conditions are not rigorously controlled an ECD can behave as mass sensitive because the electron density (and so response) is a function of carrier flow rate.[34]

## Peak Area and Solute Quantity

If peak area is proportional to solute quantity it must be so for both types of detector.

### Flow Sensitive Detectors

In a flow sensitive detector, the output signal, $S$, is proportional to the concentration, $c$, of solute in mobile phase

*i.e.* $c = KS$ where $K$ is a constant $\hspace{2cm}$ (56)

integrating this from the start to the end of the peak:

$$\int_{s}^{e} C \, dt = K \int_{s}^{e} S \, dt \hspace{2cm} (57)$$

$$= KA \hspace{2cm} (58)$$

where $A$ is the peak area.

The left hand side of Equation 57 can be transformed:

$$\int_s^e c\frac{dt}{dV}dV = KA \tag{59}$$

or

$$\int_s^e c\frac{1}{\dot{V}}dV = KA \tag{60}$$

where $\dot{V}$ is the flow rate of mobile phase. As long as it is constant, it can be taken out of the integration and moved to the right hand side of Equation 60:

$$\int_s^e c\,dV = KA\dot{V} \tag{61}$$

what remains on the left hand side is simply the quantity of solute, $Q$ in the eluting zone, *i.e.*

$$Q = KA\dot{V} \tag{62}$$

For a flow sensitive detector, peak area is proportional to solute quantity provided that flow rate is constant. The majority of LC detectors and the TCD are flow sensitive.

It is therefore very important to maintain precise flow control when using flow sensitive detectors. On the other hand, peak area and sensitivity can be increased by reducing the flow rate though the analysis will take longer (Figure 1.25). If the flow rate is reduced too far, column efficiency will suffer. Bakalyar[35] studied the effect of flow and solvent composition stability on area and height measurements made on a UV detector. The results are summarized in Table 1.4.

**Table 1.4** *Effects of Chromatographic Changes on Data from UV Detector*[35]

|  | Flow rate −1% | Composition −1% B | Temperature −1 °C |
|---|---|---|---|
| Area | +1% | 0 to ±10% | 0 to ±1% |
| Height | < +0.3% | −1 to −10% | 0 to ±1% |

**Mass Sensitive Detectors**

The output signal from a mass sensitive detector is proportional to the rate at which the solute is consumed by the detector reaction, or, as the reaction rate must be faster than the flow rate for quantitation to be possible, to the rate at which solute is delivered to the detector.

$$i.e. \frac{dQ}{dt} = KS \tag{63}$$

integrating this from the start to the end of the peak:

$$\int_s^e dQ = K \int_s^e S.dt \tag{64}$$

$$\text{or } Q = KA \tag{65}$$

For a mass sensitive detector, peak area is proportional to solute quantity but independent of flow rate, which conveniently removes flow rate as a potential source of experimental error.

However, peak shape varies with flow rate. If the flow rate is high, a given solute quantity will produce a tall slim peak. If the flow rate is low, the same quantity will produce a low broad peak of longer retention. In both cases peak area will be the same (Figure 1.25b).

**Flow Sensitive**

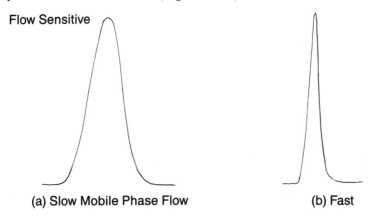

(a) Slow Mobile Phase Flow          (b) Fast

The area is smaller for the faster flow rate

**Mass Sensitive**

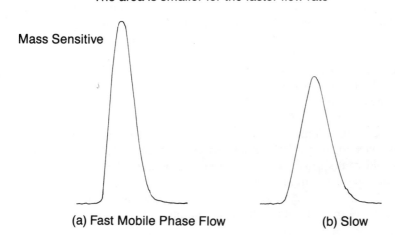

(a) Fast Mobile Phase Flow          (b) Slow

Peak shape changes but area is the same

**Figure 1.25**  *Comparison of flow sensitive and mass sensitive detectors*

Increasing the flow rate through a mass sensitive detector will speed up an analysis without loss of sensitivity, but it will invalidate calibrations of height made at the slower flow.

### Detector Overload

Detectors have a finite range of operation. FID electrometers only accept an input signal from peaks up to a certain size because the output signal is restricted to a maximum of 1 V DC. Bigger peaks are clipped (see Figure 1.26).

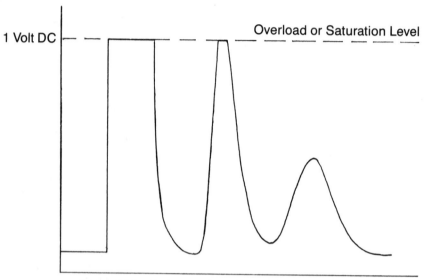

**Figure 1.26**   *Peak area above the overload level is lost*

As long as important peaks are kept on scale, which is usually the case when using a chart recorder, saturation can be spotted. However, integrators and computers also measure peaks which go off scale and saturation can be missed.

Detector cells distort peak shapes through geometric or mechanical constraints by being too big or the wrong shape.[12,36] FID flames quench and vary their response considerably when certain solutes, especially halocarbons, pass through. Non-linear thermal conductivity effects cause hydrogen peaks in helium carrier gas to fold over when measured by TCD at critical concentrations.[37,38] Under certain conditions, helium ionization detectors fold peaks too.[39] The folding occurs because the linearity curve has a maximum or minimum in it.[40]

The effect of overloading or saturation is to make measured areas smaller than they should be by an unpredictable amount.

## Linearity

Below saturation level, there is a linear working range where detector response is proportional to solute quantity. This range is greater for peak area than peak height.[40,41] Above it there is a non-linear but maybe useful range where the relationship between solute and response is continuously changing. Finally there is saturation and no relationship.

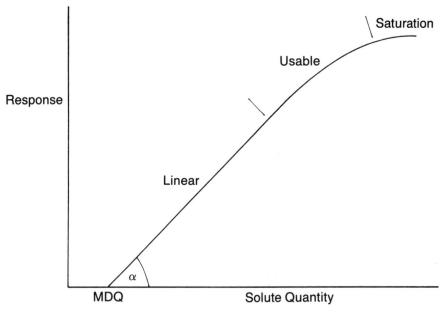

**Figure 1.27** *Linear dynamic range. MDQ = minimum detectable quantity; α = sensitivity*

In the linear range the relationship between solute quantity and detector response, output voltage $V$ is:

$$V = KRQ \tag{66}$$

where $Q$ is solute quantity, $R$ is solute response factor and $K$ is an instrument constant. MDQ is small enough to ignore.

Fowlis[42] has suggested that a better equation is:

$$V = KRQ^r \tag{67}$$

The index, $r$, puts a numerical value to the meaning of linearity; the detector can be considered to be working linearly if $r$ lies between 0.98 and 1.02.

Carr[40] generally expressed detector linearity by the polynomial:

$$V = \sum_{i=0}^{n} P_i Q^i$$

$$= P_0 + P_1 Q + \sum_{i=2}^{n} P_i Q^i \tag{68}$$

In this form $P_0$ is the minimum detectable quantity, $P_1Q$ is the linear

term and $\sum_{i=2}^{n} P_i Q^i$ is the sum of all the non-linear terms among which

the quadratic is the dominant term.

It is better to determine by experimentation where the limits of linearity are rather than trust manufacturer's specifications. Carr also found that in UV detectors, non-linearity can occur at unexpectedly low concentrations, less than 1 millimole, and it may not be detectable by inspection.

Ultimately there is little use in judging the performance of a detector and its electronics separately except when trouble shooting. The linear dynamic range of a detector is determined by both.

**Detector Non-Linearity** normally affects the top of the peak but not the bottom which remains in the linear region. It reduces peak height and peak area but does not affect retention time or peak skew unless the detector actually saturates or the peak is already asymmetric, in which case it might shift the peak maximum further. The fractional loss in area due to detector non-linearity is always less than fractional loss in height because area is lost from the top of the peak where it matters least.

**Column Non-Linearity** affects the peak from top to bottom. It causes peak skew which affects peak shape and therefore height and retention time, but peak area remains unchanged.

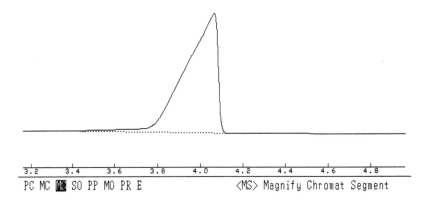

**Figure 1.28**   *Column asymmetry*

**Electronic Noise and Drift**
As technology improves, noise and drift become less, but this only encourages analysts to measure smaller quantities and so the problem never disappears. For routine analyses, where solute quantity is not a problem, electronic component noise and drift are usually much less than transducer or column noise and can be ignored except as evidence of a malfunction.

# 11 Sources of Peak Measurement Error

Assuming the chromatograph is used within its operating range and the detector signal is an accurate representation of solute profile, integration errors are caused by:

(a) noise;
(b) baseline drift;
(c) separation of incompletely resolved peaks;
(d) peak asymmetry;
(e) incorrect use of the integrator.

Category (e) is deferred to Chapter 3.

## Noise

Noise is unwanted detector signal and can be chemical or electronic in origin. What the analyst sees is the residue left after filtering and smoothing have removed higher frequencies. It has been allowed to remain because its frequency is too close to that of known chromatographic peaks, and filters which would remove it might also remove wanted peaks.

It can be measured statically or dynamically, with the mobile phase either stationary or flowing. Static measurement measures detector noise alone; dynamic measurement includes noise from the rest of the system. It should make no contribution, but if it does it warns that the detector signal is being degraded.

### Noise and Frequency
Noise is characterized by its frequency and peaks by their width; one is the reciprocal of the other:

$$\text{Peak 'Frequency'} = \frac{1}{\text{Peak Base Width}} \tag{69}$$

The aim of instrument design engineers is always to minimize the amount of noise created (since what is not created does not have to be removed), filter as much of the created noise as possible, and finally improve the signal processing techniques to measure only the peaks.

Three kinds of chromatographic noise are recognized in ASTM E 685–79,[43] as shown in Figure 1.29.

### Short Term Noise
Defined as: random variations in detector signal whose frequency is greater than 1 cycle/minute. Ideally, short term noise is narrower than 'real peaks' and this difference can be used by integrators to recognize and disregard noise peaks. Continuing developments in capillary column technology squeeze this definition; small capillary peaks are increasingly confused with short term noise and integrators may then filter them.

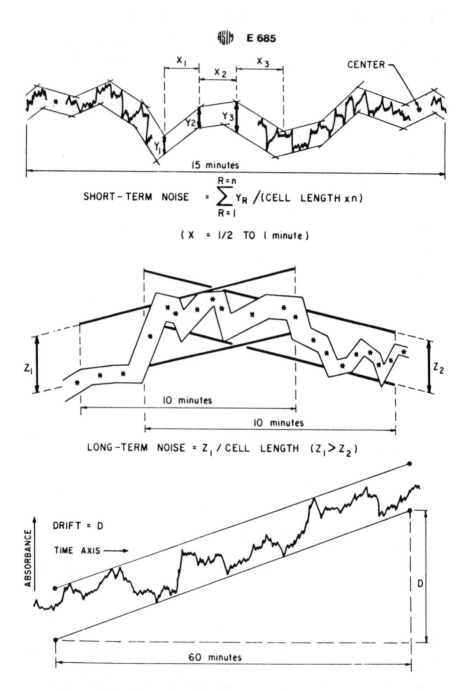

**Figure 1.29**   *Three types of noise recognized in* ASTM E 685-79
(Reproduced by kind permission of ASTM[43])

### Long Term Noise

Variations in detector signal whose frequency lies between 6 and 60 cycles/hour. It is the kind of noise which looks like peaks and may very well be late eluters from an earlier analysis trespassing into the current one.

### Drift

Drift is change in baseline position. As noise, its frequency is even less than long term noise. If it cannot be attributed to some known cause such as temperature or solvent programming, it indicates instrument instability, often due to temperature effects on the detector.

### Errors Created by Noise

Noise affects integrator measurement of peaks around those regions where the peak signal is near horizontal. It blurs the base of a peak making it difficult to locate where it starts and ends and therefore where to measure the area. Noise at the top of peaks has caused integrators to split area measurement as valley recognition is triggered among the micro-peaks. In the subsequent report these peaks are observed to have retention times within a few seconds of each other. Similar splitting can occur in valleys or cause the integrator to locate the valley in the wrong place and transfer a slice of one peak to the other. Long term noise is measured as peaks and included in the analysis report.

Noise sets a limit to the smallest quantity of solute that can be detected as a single peak. Small peaks don't simply disappear from view: electronic amplification would always bring them back. They disappear when they merge into the baseline noise and become indistinguishable from it.

### Signal to Noise Ratio: The Smallest Measurable Peak

The smallest fully resolved peak which can be measured is the smallest one which can be unambiguously distinguished from baseline noise. It is described in terms of the signal to noise ratio, $S/N$, which compares the height of a peak to the height of the surrounding noise.

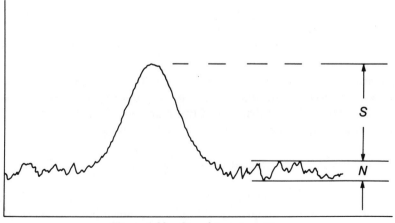

**Figure 1.30** *Signal to noise ratio*

If a solute is repeatedly analysed and diluted, it is judged that its peak can still be seen on a chromatogram when the $S/N$ ratio is as low as 1, *i.e.* the peak is as high as the noise amplitude, but in this kind of experiment the analyst knows the retention time of the peak and knows where to look.

As the $S/N$ ratio increases to 2, the peak is more clearly visible, and when it is 3 it is unambiguous though area measurement will be imprecise.

### Limits of Detection and Quantitation

ACS guidelines[44] published in 1980 define two measures (Figure 1.31):

(1) Limit of detection, LOD, at $S/N = 3$ defines the smallest peak that can be confidently judged to be a peak.
(2) Limit of quantitation, LOQ, at $S/N = 10$ defines the smallest peak whose area can be measured with acceptable precision.

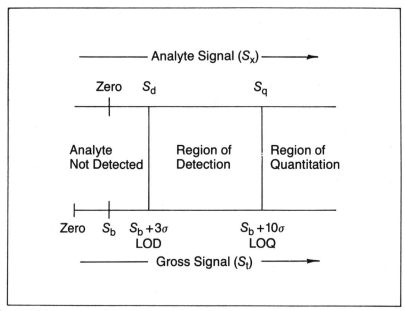

**Figure 1.31**  *Limits of detection and quantitation*
(Data reproduced with kind permission from *Anal. Chem.*, 1980, **52**, 2242)

Quantitation below $S/N = 10$ is not advised for integrators in particular because of the risk of detecting noise as peaks and of losing peaks in noise. Kaiser calculates a 7% probability of this happening at $S/N = 3$.[45]

In these guidelines signal is measured above the noise, so that:

$$\text{Signal/Noise ratio} = \frac{S}{N} \qquad \text{in Figure 1.30}$$

Earlier definitions have used $(S+N)/N$, but when no peak is present ($S=0$), the ratio has the value 1 implying that there is as much peak signal as noise and that solute is still there.

**A 'Good Baseline'**

Studies of the third and fourth peak moments, skew and excess, which are very sensitive to noise around the limits of integration, have brought some workers[25,46] to judge that the accuracy of area and retention measurement is acceptable when $S/N$ is greater than 30 : 1, but a 'good baseline' is where the $S/N$ ratio exceeds 100 because only at this level do the errors in measurement of variance, skew and excess become reasonably small.[25]

## Errors Created By Baseline Drift

Baseline drift causes precision errors when it is not constant over a series of analyses, and inaccuracies in peak measurement even when it is repeatable.

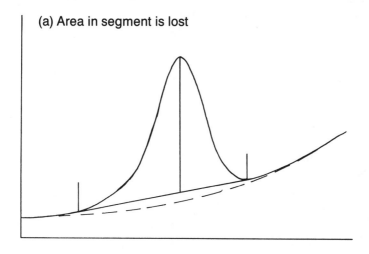

**(a) Area in segment is lost**

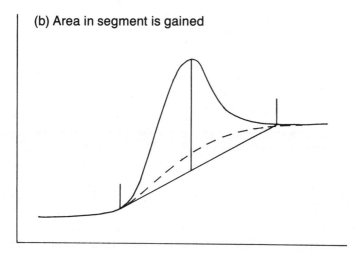

**(b) Area in segment is gained**

**Figure 1.32** *Linear constructed baselines are not necessarily accurate*

### Drifting Baseline and Peak Measurement

The baseline constructed for peak measurement is a straight line drawn beneath the peak. If real baseline is defined as the detector signal in the absence of the peak, then the dashed curves in Figure 1.32 are better approximations. Peak areas and heights will be measured greater or smaller than they should be depending on whether the true baseline is convex or concave.

The magnitude of the error depends on the curvature of the true baseline, the position of the peak on the curve, its width and peak overlap, which extends the required length of constructed baseline. These factors can conspire at different times to make the error serious or negligible. Single capillary peaks on a solvent tail are measured with good accuracy; groups of broad peaks on a steeply rising baseline near the end of the analysis may be measured with very poor accuracy (Figure 1.33).

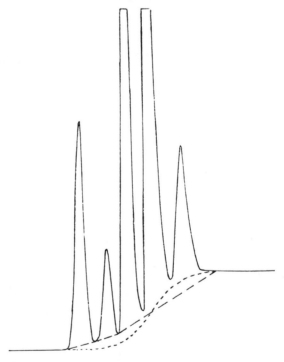

**Figure 1.33**   *The dotted baseline represents the true baseline; the dashed baseline is the one placed by the integrator*

### Baseline Drift and Retention Time

The retention time of a peak measured on a positive baseline slope is greater than it should be because the gradient of the baseline adds to the peak top and delays the maximum. A new peak maximum is created at the point where the negative gradient of the falling peak slope is equal in (absolute) value to the positive gradient of the baseline (Figure 1.34).

Similarly, the measured retention time of a peak on a negatively sloping baseline is always less than the true retention time.

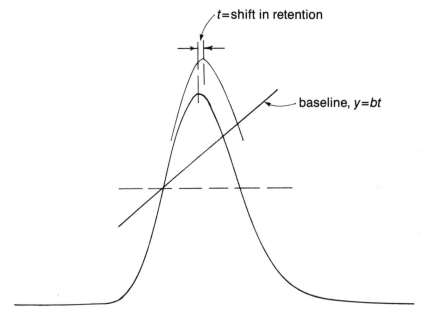

**Figure 1.34** *Effect of baseline drift on retention time*

Near the peak maximum, peaks are parabolic in shape and the height, $y$, of a point near the maximum is given by:

$$y = H\left(1 - \frac{t^2}{2\sigma^2}\right) \qquad \text{see (26)}$$

where $H$ is the peak height

$t$ is time on a scale where retention time $= 0$

and $\sigma^2$ is the peak variance

For simplicity, in the parabolic region baseline curvature and differences in response are neglected. Positive or negative baselines are represented by:

$$y' = bt \qquad (70)$$

where $b$ is the baseline gradient

The total signal, $S$, is:

$$S = H + bt - \frac{Ht^2}{2\sigma^2} \qquad (71)$$

and this function has zero slope where:

$$t = \pm \frac{b\sigma^2}{H} \tag{72}$$

Equation 72 represents the shift in observed retention time. The shift is directly proportional to baseline gradient. On a flat baseline ($b = 0$) the peak maximum occurs at $t = 0$. For broad peaks, $\sigma^2$ is large; for narrow peaks it is small and the shift in retention is proportional to peak variance.

Equation 72 also shows the shift in retention time to be inversely proportional to peak height. For a peak of given area, the taller it is, the narrower it is and the smaller the retention shift.

### Reduced Detector Operating Range
If detector drift is not checked, the baseline can drift up towards the upper limit of detector operation. This will reduce the linear working range of the detector and bring about premature saturation and loss of peak area.

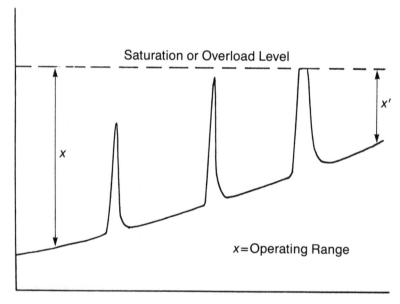

**Figure 1.35** *An upwardly drifting baseline restricts operating range*

A negative drifting baseline can equally drift below the lower operating limit of an integrator so that peak base area is lost, and only that part of peak which rises into the operating range is measured.

Careless use of detector back-off control can move the baseline towards either operating limit.

Baselines must be kept close to ground, or virtual ground, by regular 'zeroing' of the detector signal.

**Some preliminary conclusions**

(1) Analyses are more accurate on flat baselines than on sloping (curved) baselines, whatever the cause of the slope.
(2) Narrow peaks on sloping baselines are subject to less error in area and height measurement than broad peaks because the degree of departure between the constructed baseline and the 'true' curved baseline will be less for narrow peaks.
(3) The error in measuring retention time of a peak on a sloping baseline increases with the baseline slope but is less for tall narrow peaks than for low broad peaks.

## Errors of Incomplete Peak Resolution

Integrators and computers separate fused peaks by dropping perpendiculars from the valley between them or, if one peak is much smaller than the other and located on its tail, by skimming a tangent below it (Figure 1.36).

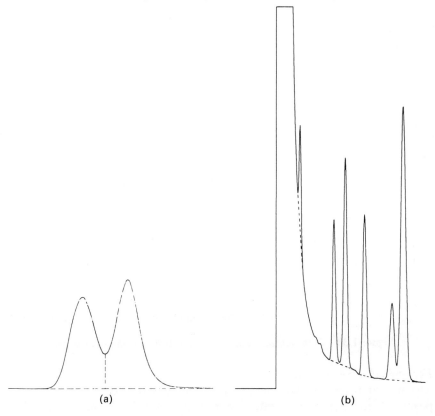

(a)                                    (b)

**Figure 1.36** *Separation of fused peaks*

For manual measurement of overlapping peaks, triangulation is used (Figure 1.37).

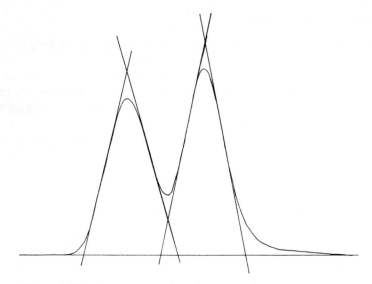

**Figure 1.37**   *Measurement of overlapping peaks by triangulation*

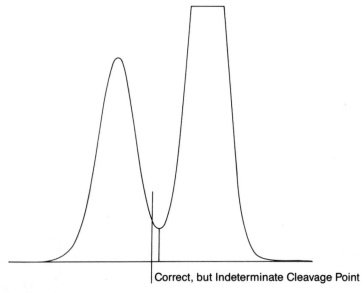

Correct, but Indeterminate Cleavage Point

**Figure 1.38**   *Perpendicular separation over-estimates the smaller peak*

### Perpendicular Separation

The use of perpendiculars or triangulation to separate two overlapping peaks will give rise to inaccurate area measurements[33,47,48] unless the peaks are:

(1) same height and width and are symmetrical in shape;

(2) valley is no more than 5% peak height;

(3) baseline is flat;

(4) noise does not obscure peak starts, ends and valley locations.

These are exceptional circumstances.

If the peaks are symmetrical but unequal in height, perpendicular separation over-estimates the smaller area at the expense of the larger one (Figure 1.38).

If, in addition, the peaks are asymmetric (tailing), there is a great inequality in the contributions of each peak to the other which can result in a gross over-estimate of the area of the second if a perpendicular is used (Figure 1.39).

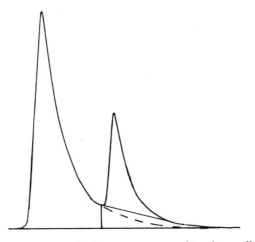

**Figure 1.39** *It may be less inaccurate to skim the smaller peak*

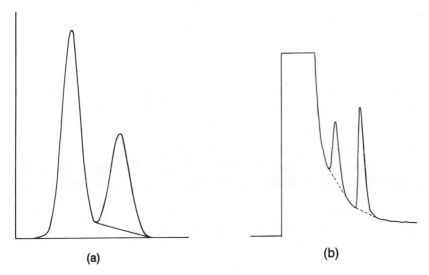

(a)

(b)

**Figure 1.40** *Tangent skimming for smaller peaks: (a) is not very accurate; (b) is accurate*

### Tangent Skim Errors

The use of tangent skimming carries the error to the opposite extreme. Tangent skimming under-estimates the area of the smaller peak unless it is very much smaller and narrower than the peak from which it is skimmed (Figure 1.40).

### Errors of Perpendicular/Tangent Transition

Another kind of error is introduced when integrators measure fused peaks of such marginal size that in one analysis separation is by perpendicular but in another a tangent is skimmed beneath the smaller peak (Figure 1.41).

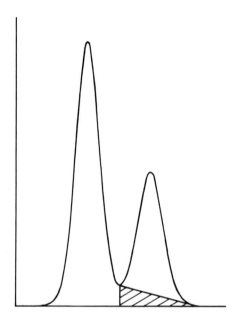

**Figure 1.41** *The change from perpendicular to tangent transfers the shaded area to the large peak*

To overcome this, integrators can force separation by one method exclusively. This increases precision but does not necessarily improve the accuracy of measurement.

### Effect of Overlap on Area Measurement Accuracy

The error in peak area introduced by separating overlapping peaks by perpendiculars and triangulation has been studied quantitatively by Westerberg.[47]

The error caused by each method was studied for two symmetrical peaks of equal width (Figure 1.42) as a function of relative peak size, $R$, and separation, $\delta$, where:

$$R = \frac{H_2}{H_1} \qquad \text{and} \qquad \delta = \frac{t_2 - t_1}{\sqrt{(2.\sigma)}} \tag{73}$$

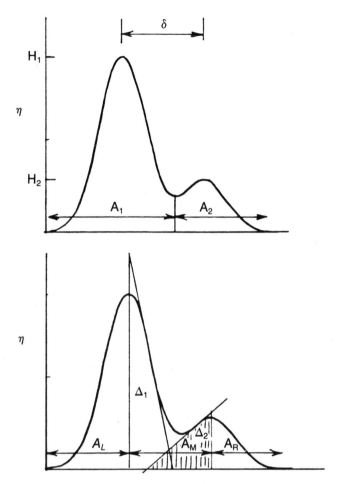

**Figure 1.42** *Comparison of overlapping peak area measurement by perpendicular and triangulation*
(Redrawn with kind permission from *Anal. Chem.*, 1969, **41**, 1770)

The smaller peak is over-estimated by both methods and the measurement error increases as peak separation is reduced and the peak height ratio increases (Figure 1.43). Conversely, the error disappears when resolution is complete.

Similar studies by Vandeginste[49] with symmetrical peaks of unequal width drew similar conclusions.

The shoulder limit in Figure 1.43 is where the valley disappears and becomes a point of inflection. The detectability limit is where the two peaks appear to be one. As the peaks merge, the fact that two peaks are present remains clearly visible for the least resolution when the peaks have a relative height of $R = 0.446$. When the peaks are allowed different widths, this point of maximum detectability occurs at smaller values of $R$ as the second peak broadens.

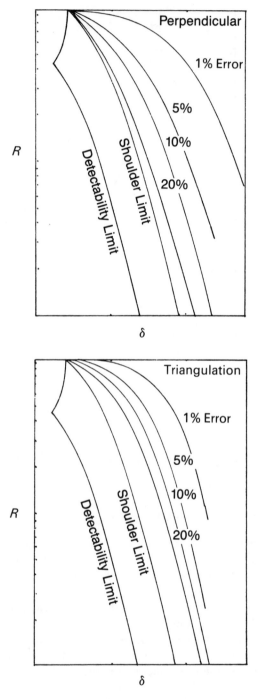

**Figure 1.43**   *Comparison of error using perpendicular or triangulation*
(Data reproduced with kind permission from *Anal. Chem.*, 1969, **41**, 1770)

Comparing techniques, perpendicular separation is less sensitive to over-lap errors than triangulation because the error contours on the perpen-dicular graph are more widely spaced than on the triangulation graph.

Proksch[48] prepared tables of correction coefficients to compensate for the errors when pairs of variously sized, symmetrical peaks are separated this way, but they are only accurate when the smaller peak elutes first.[50]

Using computer generated symmetrical and unsymmetrical peaks, Foley[51] has shown that by using perpendicular separation to separate various combinations of overlapping peaks, errors up to 50% in area measurement are not uncommon, and in staged examples errors up to 200% are achievable.

All of these attempts to correct the error of perpendicular separation limit themselves to the two peak example. With three overlapping peaks the situation is much more complex. The triplet has always been an early discard in deconvolution studies.

Changing from perpendicular to tangent skim reduces the extreme errors when the peak height ratio becomes too small (less than 10:1 approxi-mately[65]), but the point of transition from perpendicular to tangent skim is arbitrary.

The only sensible conclusion to be drawn is that errors arising from overlap should be solved by chromatography. Integrators will generate a highly precise but inaccurate set of results for all the foregoing examples.

### Height of a Fused Peak

If part of either peak is under the peak maximum of the other, it lifts it and creates an error in the peak height measurement. This error is greatest when a small peak sits on the tail of a larger, asymmetric one (Figure 1.44).

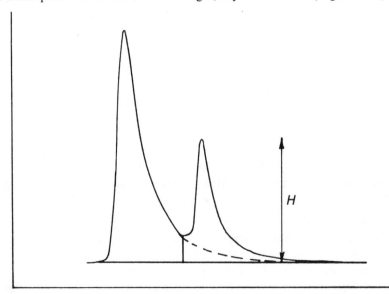

**Figure 1.44** *The second peak is lifted by the tail of the first and is measured too high*

A peak between two other peaks is lifted from both sides making height measurement meaningless. However, it might still have some marginal value for quality control comparisons if the 'floating' height can be used as a diagnostic of instrument stability.

**Overlapping Peaks on Sloping Baselines**

Sloping baselines compound the errors of overlap in two specific ways:

**(1) Shifting the valley position.** If a valley is thought of as an inverted peak top, then just as retention time is shifted by sloping baseline, so is the valley but in the opposite direction. A positive baseline will bring the valley forward in time, a negative baseline will delay it (Figure 1.45).

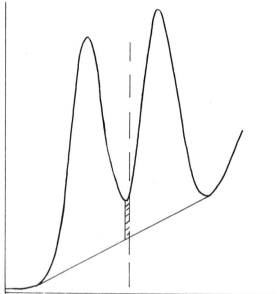

**Figure 1.45**    *The dashed line shows where the valley perpendicular would have been on a horizontal baseline. The shaded area is transferred from the first peak to the second*

Whether this shift introduces serious error or not depends on baseline gradient, peak width and, additionally, on the degree of peak overlap. For narrow peaks and shallow gradients the errors can be ignored. Larger shifts will occur for broad peaks on gradients (up to about 6 or 7%[66]). If resolution is poor and the valley is high the effect of the shift will be to shave a tall slice off one peak and add it to the other. With good separation the valley is low and the transferred area will be small (Figure 1.46).

**(2) Position of peaks on the slope.** The curvature of a peak tail is not uniform, and in consequence when it acts as a baseline for groups of small peaks the accuracy of peak measurements depends on the position of the smaller peaks on the tail. The error in these measurements will be greatest where the departure between tangent baseline and tail curve is worst, where the curvature of the tail is greatest.

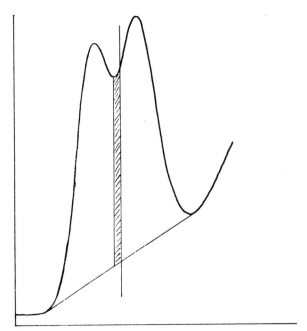

**Figure 1.46**  *The transferred area is significant when the valley is high and the gradient large*

Errors are least when the rider peaks are narrow and situated at the end of the tail.

The slope of the tail, like baseline slope, affects the position of the valley and peak maximum. Integrators skim small peaks from the valley position,[52] not the tangent position preceding it, and the error this causes will be greatest when the peak is high up the tail (Figure 1.47). Such errors could be avoided by a simple change to the integrator software.

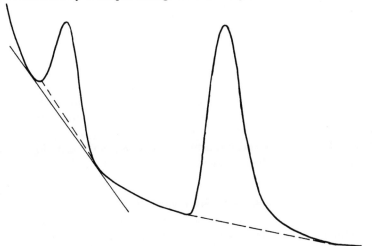

**Figure 1.47**  *Tangent is skimmed from previous valley, not tangent point before valley. Therefore, error varies with position on tail*

## Multiple Fused Peaks

Given the errors inherent in the measurement of fused pairs, larger groups (even three peaks) clearly offer greater opportunity to compound errors. The central peaks in particular are disturbed on both sides. It is therefore very important to configure quantitative analyses to minimize baseline gradients and peak overlap. If analysts can choose their analyses—as with prepared solutions—and arrange to have peaks of similar size it will reduce errors where overlap is unavoidable.

Commonly, the chromatographer is confronted by a complex chromatogram on a shifting baseline with little choice but to accept it and measure it as it is.

Calibration reduces the errors by adding compensation to the response factors, but results should never be accepted unquestioningly.

Overlapping peaks as shown in Figure 1.48 are not an unusual occurrence, and with groups like these the analyst cannot even be sure how many peaks are present[53,54] let alone measure their areas or heights.

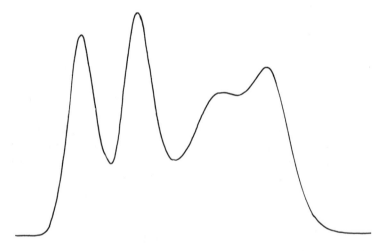

**Figure 1.48**  *Looks like four peaks—but it is only a guess*

In these circumstances, the object of quantitation is to produce comparable sets of results. Two solutions that produce the same chromatograms and have the same measured composition are judged to be the same material. Repeatability and not absolute accuracy is the criterion to accept. It is a substitute that has often proved fallible but survives for want of a better alternative.

## Mathematical Deconvolution of Overlapping Peaks

Attempts[55,56,57,58] have been made to deconvolute overlapping peaks by mathematical techniques and so avoid the need to use perpendiculars and tangents. All studies have required peaks to fit a mathematical model peak shape such as Gaussian, and all have failed for two basic reasons:

(1) Peaks will not conform to a single mathematical model. Even within the same chromatogram a range of models might be necessary.

(2) There is simply not enough information contained within a single channel detector signal to allow numbers of peaks of indeterminate shape to be extracted from a group.[59] Curve fitting procedures only work when the number of peaks present is known.[49]

Perpendiculars and tangents remain in use because deconvolution studies have failed to deliver a better method. They are straight lines and can at least be drawn with minimal subjective judgement. Non-linear separation of tailing peaks using, for example, exponential curves has to be justified and the curve correctly located, and though an integrator could be given rules to do this, the analyst would have no easy way to test the accuracy of such a technique.

## Errors from Peak Asymmetry

Skew is peak asymmetry about a vertical axis. It is positive if the peak tails and negative if the peak 'fronts'. It occurs round the base of a peak much more than the top and can be thought of as unsymmetrical base broadening made possible by transfer of area from the top of the peak to the bottom. In fact the top of the peak often remains symmetrical while the base skews severely.[60]

Asymmetry caused by pre-column broadening and slow equilibration between solute and column will dissipate as retention increases. This is sometimes used as a visual inspection for injection proficiency.

Asymmetry caused by non-linear adsorption isotherms persists at all retentions.

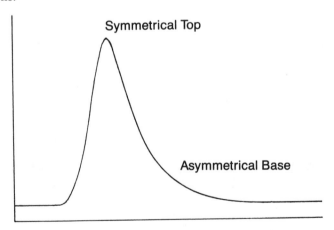

**Figure 1.49** *The top of an asymmetric peak can be quite symmetrical*

Asymmetry contributes to a number of direct and indirect errors in peak measurement.

### Asymmetry and Peak Tailing

Tailing makes it harder for the integrator to track the end of the peak as it merges into the baseline noise. The end limit of integration is invariably located too soon, area is lost, and the measured peak area is too small (Figure 1.50).

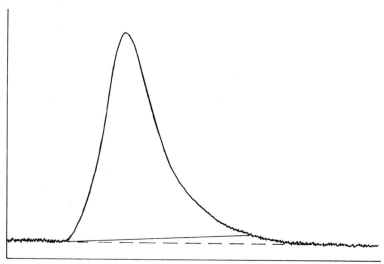

**Figure 1.50**   *Asymmetry causes peak tail to be lost under the noise*

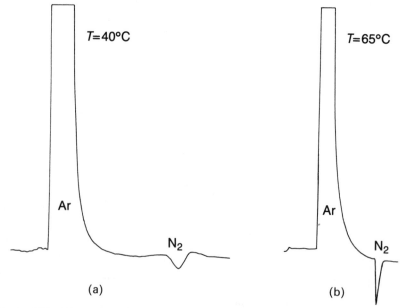

**Figure 1.51**   *Trace* $N_2$ *peak in argon on Molecular Sieve 5A columns.* $H_2$ *carrier gas and TCD*

Tailing can persist after the recorder or plotter has ceased to show it. Figure 1.51 shows a trace of about 100 ppm of nitrogen in argon measured on a TCD using hydrogen carrier gas. Both peaks should be positive because the thermal conductivities of both are less than that of the carrier gas, yet the nitrogen peak is inverted.

In spite of appearances, argon has not finished eluting and traces are still inside the TCD cell when the nitrogen emerges. Since nitrogen has a higher thermal conductivity than argon there is a small rise in the heat loss from the detector filament which gives rise to the negative peak. Temperature control of the resolution and therefore of the concentration of nitrogen in argon allows the tail to be monitored.

At 40 °C any integrator would detect the end of the argon peak before the start of the nitrogen and would therefore fail to measure the whole of the argon tail, but it is clear from Figure 1.51(a) that the argon tail is not visible to be measured.

Jover and Juhasz[61] estimated how much area was being lost by fitting tail ordinates to an exponential curve and estimating the tail area by extrapolation of the curve. They showed that the extrapolated peak area was more than 4% larger than the conventional measurement when the peak end was judged by eye.

### Asymmetry and Base Broadening

Asymmetry broadens peak bases and creates a loss of resolution as defined by equation 4 on page 6, because of increased base width. At the same time it can appear that there has been an increase in resolution because peak top narrowing accompanies base broadening and so the valley height falls (Figure 1.52). Asymmetry increases the overlap between two peaks of fixed separation and creates or increases the overlap problems described earlier.

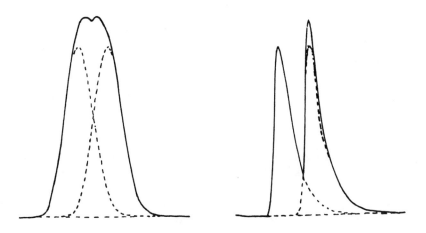

**Figure 1.52** *Paradoxically, asymmetry can appear to improve resolution*
(Reproduced by kind permission from *J. Chromatogr. Sci.*, 1977, **15**, 303)

When a peak tail disappears under the following peak, perpendicular separation transfers the tail into the second peak (Figure 1.53).

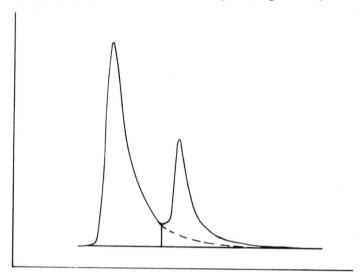

**Figure 1.53**   *Asymmetry causes peak tail area to be lost under, and added to, the next peak*

### Asymmetry and Manual Peak Measurement

Commonly used manual peak measurements are based on the Gaussian peak model. Formulae for triangulation and peak height × width calculations inherently assume symmetry. If this is no longer true, manual measurements of peak area (and height) cannot be assumed to be proportional to solute quantity. Other measurement techniques which tolerate asymmetry should be adopted.

### Unequal Asymmetry

Problems of asymmetry might be avoided if all peaks were asymmetrical in exactly the same way: then the various formulae could be modified by adding a 'shape factor'. For example, the area of a single symmetrical peak is:

$$\text{Area} = k_i . H_i . w_{0.5,i} \tag{74}$$

in which $H_i$ is the peak height, $w_{0.5}$ is the peak width at half height and $k_i = 1/0.939$. If the peak is not symmetrical $k_i$ will have a different constant value, but as long as all peaks are the same shape they will all have the same $k$ value which will cancel when peaks are compared:

$$\frac{A_1}{A_2} = \frac{kH_1 w_{0.5,1}}{kH_2 w_{0.5,2}} = \frac{H_1 w_{0.5,1}}{H_2 w_{0.5,2}} \tag{75}$$

or normalized:

$$A\% = \frac{kH_iw_{0.5,i}}{\sum(kH_iw_{0.5,i})} \times 100 \tag{76}$$

$$= \frac{kH_iw_{0.5,i}}{k\sum(H_iw_{0.5,i})} \times 100 \tag{77}$$

$$= \frac{H_iw_{0.5,i}}{\sum(H_iw_{0.5,i})} \times 100 \tag{78}$$

The real problem of asymmetry is that it is not uniform throughout the chromatogram; it can vary with peak size and from one peak to the next. A different constant, $k_i$, is required for each peak shape and the above simplifications cannot be made.

## Peak Area vs Peak Height

It has been observed that peak height is sometimes a more precise measure of solute quantity than area when the accuracy of each measurement is comparable. The question has been posed why this should be so. Since precision reflects the conduct of the analysis, the question can be rephrased 'Why should either be more precise than the other?'

Area is the true measure of solute quantity; height is a substitute which in the days of manual measurement conveniently speeded up measurements and dispensed with much of the drudgery. Being a simpler measure requiring fewer manual constructions to extract, it was more precise provided that good resolution and peak shape were achieved.[62] If it was less precise, it was concluded that experimental control should be improved, not that area was a better measure.

Integrators now measure height and area in similar ways. Differences in precision between each measurement mean that some aspect of the analysis procedure is not sufficiently controlled and area is a more sensitive indicator of this than peak height.

Area and height have been shown to suffer the shortcomings of chromatography in different ways. Selecting which one to use for quantitation depends on the prevailing chromatography. It is a balancing act:

**Noise** affects the base of a peak and causes area and height to be lost at the base but it affects area more than height. Height is only affected by inaccurate placement of the constructed baseline but in this case the fractional loss of area is always greater than the fractional loss in height.

**Peak Asymmetry.** Asymmetry (skew) distorts peak shape and height but not area. Area is always the more accurate measure for asymmetric peaks. When asymmetry varies from peak to peak, height cannot be used without careful calibration. It is worth reflecting that chromatograms which show no asymmetry are very rare.

**Peak Overlap.** No practical, accurate way to separate and measure unresolved peaks has yet been developed, so perpendiculars and tangents continue to be used.

When overlapping peaks are separated by a perpendicular, it is the smaller peak which suffers the large error, and the error increases with the size ratio until tangent skimming comes into effect and reverses the trend.

Height and area measurements are perturbed though height is not affected until one peak overlaps the maximum of the other; until then height is more accurate. Unfortunately, judging whether this degree of overlap exists is pure guesswork; since height and area measurements are not normally mixed in one analysis, area is preferred.

**Baseline Drift.** Construction of a linear baseline beneath a peak instead of a more accurate, but indeterminate, curved baseline affects both area and height. Baselines concern the base of the peak and area is distorted more. The area error is approximately equal to half of that caused by inaccurate placement of the baseline due to noise—which still leaves the fractional loss in area greater than the fractional loss in height.

**Baseline Disturbances.** Baseline disturbances, especially those which send the signal negative near peaks of interest, can cause large baseline placement errors which affect area more than height.

**Detector Non-Linearity and Saturation.** Height and area are reduced but the fractional loss in area is always less than the fractional loss in height because the loss is made at the top of the peak where it matters less for area.

In addition, the linear dynamic range of a detector is less for height than for area so the onset of height error starts at lower concentrations.

**Detector Type.** If mobile phase flow rate is imprecise, area will be the better measure of peaks produced by mass sensitive detectors like the FID, and height will be more accurate when using flow sensitive detectors such as UV. Most flow sensitive detectors are used for LC, however, and gradient solvent composition also must be considered: if solvent composition is not precisely controlled then even for flow sensitive detectors, area is more accurate.

A similar consideration applies to TCDs and temperature programming.

**Noise Filtration.** Instrument time constants must be fast enough to allow the fastest eluting peaks to pass, and RC filters ought to have the same effect (*i.e.* zero) on all fast peaks.

The analyst who is measuring very fast peaks cannot always be sure that this is true. Early eluters may be distorted in a way that is slightly different from later, broader peaks, putting height determination in doubt. If sampling frequency is not high enough, integrator smoothing techniques based on Savitsky–Golay have been shown to distort both,[63] but height more than area. Since the analyst has no quick way to check that smoothing distortion is absent, area must be the preferred measure.

Provided that noise is not excessive and the integrator has drawn the baseline in the correct place, it is better to measure area than height. The only time that height is likely to be a more accurate measure is when $S/N$ ratio is poor ($<30$), or what creates a similar effect, when very small peaks litter the baseline close to large peaks of interest and narrow the limits of integration.

If height is a more precise measure than area when the quality of the chromatography is good, it is a warning that the integrator is not programmed correctly. Adjustments should be made to any or all of the principal parameters: sampling frequency, slope sensitivity and baseline range parameters (see Chapter 3).

An ASTM survey of 50 laboratories in 1984[64] set out to compare the precision of height and area meaurements. Similar samples were analysed by LC using integrators to measure area and a mix of integrators and manual techniques to measure height. It was found that when the chromatography was uncomplicated, *i.e.* no small peaks, and all peaks were nearly symmetrical and well resolved, area gave marginally better accuracy and was more precise than height.

When the analysis contained perturbations due to overlap and detector non-linearity, accuracy and precision began to suffer. If a peak was perturbed by a nearby baseline disturbance, the accuracy of the peak measurement did not suffer but height was three times more precise than area. Detector non-linearity affected large peaks, which were under-estimated. Small rider peaks were slightly over-estimated. For both of these types of peak, area was both more accurate and more precise.

The surveyors were forced to conclude, however, that the report lacked detailed information about the integrator parameter values used for the various measurements. The accuracy of the results could be interpreted in terms of the factors described above, but variations in column performance were evident and it was not clear whether the chromatograms had been processed in identical fashion, or whether differences in integration had influenced precision. Comparisons were ultimately limited by the different ways the integrators were used.

Integrators and computers should always be more precise than manual methods of peak measurement. Manual methods can be more accurate because analysts are unlikely to err in judgement. They will be consistent when drawing baselines, selecting tangents or perpendiculars for measuring overlapping peaks, and are better able to see through the effects of noise and to disregard stray peaks.

Integrators and computers will measure every peak they are not instructed to ignore, and may construct highly displaced baselines if there is a signal disturbance near peaks of interest.

The great advantage of integrators and computers over manual measurement is their speed and convenience and for these reasons analysts will persevere with the shortcomings of their measuring techniques, short-comings which can be mitigated by good chromatography—which is how it should be.

# 12 References

1. J. J. Kirkland, W. W. Yau, H. J. Stoklosa, and C. H. Dilks Jr., *J. Chromatogr. Sci.*, 1977, **15**, 303.
2. K. Pearson, 'Methods of Statistical Analysis' 2nd Edn., John Wiley and Sons, New York, 1952, p. 33.
3. A. B. Littlewood, 'Gas Chromatography Principles, Techniques and Applications', 2nd Edn., Academic Press, Inc., New York, 1970.
4. J. P. Foley, *Anal. Chem.*, 1987, **59**, 1984.
5. D. J. Anderson and R. W. Walters, *J. Chromatogr. Sci.*, 1984, **22**, 353.
6. F. D. A. Compliance Policy Guide 7124.08, 12/1/1982.
7. A. Klinkenberg and F. Sjenitzer, *Chem. Eng. Sci.*, 1956, **5**, 258.
8. C. J. Brookes, I. G. Batteley, and S. M. Loxton, 'Fundamentals of Mathematics and Statistics', John Wiley and Sons, New York, 1979.
9. CRC, 'Standard Mathematical Tables' 26th Edn., CRC Press Inc., Boca Raton, Florida.
10. A. B. Littlewood, *Z. Anal. Chem.*, 1968, **236**, 39.
11. J. P. Foley and J. G. Dorsey, *J. Chromatogr. Sci.*, 1984, **22**, 40.
12. J. C. Sternberg, 'Advances in Chromatography', Vol. 2, Marcel Dekker, New York, 1966.
13. P. T. Kissinger, L. J. Felice, D. J. Miner, C. R. Reddy, and R. E. Shoup, 'Contemporary Topics in Analytical and Clinical Chem.' Vol. 2, Plenum Press, New York, 1978.
14. V. Maynard and E. Grushka, *Anal. Chem.*, 1972, **44**, 1427.
15. R. E. Pauls and L. B. Rogers, *Anal. Chem.*, 1977, **49**, 625.
16. E. Grushka, *Anal. Chem.*, 1972, **44**, 1733.
17. R. E. Pauls and L. B. Rogers, *Sep. Sci. Tech.*, 1977, **12**, 395.
18. J. P. Foley and J. G. Dorsey, *Anal. Chem.*, 1983, **55**, 730.
19. D. A. Mcquarrie, *J. Chem. Phys.*, 1963, **38**, 437.
20. O. Grubner, 'Advances in Chromatography', Vol. 6, Marcel Dekker, New York, 1968.
21. O. Grubner, A. Zikanova, and M. Ralik, *J. Chromatogr.*, 1967, **28**, 209.
22. E. Grushka, M. N. Myers, P. D. Schettler, and J. C. Giddings, *Anal. Chem.*, 1969, **41**, 889.
23. S. N. Chesler and S. P. Cram, *Anal. Chem.*, 1971, **43**, 1922.
24. W. W. Yau, *Anal. Chem.*, 1977, **49**, 395.
25. T. Petticlerc and G. Guiochon, *J. Chromatogr. Sci.*, 1976, **14**, 531.
26. L. Condal-Bosch, *J. Chem. Educ.*, 1964, **41**, A235.
27. J. Doehl and T. Greibokk, *J. Chromatogr. Sci.*, 1987, **25**, 99.
28. S. D. Frans, M. L. McConnell, and J. M. Harris, *Anal. Chem.*, 1985, **57**, 1552.
29. L. R. Snyder and J. J. Kirkland, 'Introduction to Modern Liquid Chromatography', 2nd Edn., Wiley-Interscience, New York, 1979.
30. V. R. Meyer, *J. Chromatogr.*, 1985, **334**, 197.
31. I. Halasz, *Anal. Chem.*, 1964, **36**, 1428.

32. S. R. Bakalyar and R. A. Henry, *J. Chromatogr.*, 1976, **126**, 327.
33. W. Kipiniak, *J. Chromatogr. Sci.*, 1981, **19**, 332.
34. G. Guiochon and C. L. Guillemin, 'Quantitative Gas Chromatography for Laboratory Analyses and Process Control', *J. Chromatogr. Library*, 1988, **42**.
35. S. R. Bakalyar and B. Spruce, Rheodyne: Technical Notes 5, Dec. 1983.
36. H. Bruderreck, W. Schneider, and I. Halasz, *Anal. Chem.*, 1964, **36**, 461.
37. C. J. Cowper and A. J. DeRose, 'The Analysis of Gases', Pergamon Series in Analytical Chemistry, Vol. 7, Pergamon Press, Oxford, 1983.
38. J. G. Keppler, G. Dijkstra, and J. A. Schols, 'Vapour Phase Chromatography, Proc. of 1st Symposium', London, May 1956, Academic Press, 1957.
39. P. Deng and F. Andrews, *J. Chromatogr.*, 1985, **349**, 415.
40. P. W. Carr, *Anal. Chem.*, 1980, **52**, 1746.
41. L. M. McDowell, W. E. Barber, and P. W. Carr, *Anal. Chem.*, 1981, **53**, 1373.
42. I. A. Fowlis and R. P. W. Scott, *J. Chromatogr.*, 1963, **11**, 1.
43. ASTM E 685-79, 1st Edn., Philadelphia PA 190103, USA, 1981.
44. D. MacDougall (Chairman) *et al.*, *Anal. Chem.*, 1980, **52**, 2242.
45. H. H. Kaiser, *Anal. Chem.*, 1970, **42**, 26A.
46. D. T. Rossi, *J. Chromatogr. Sci.*, 1988, **26**, 101.
47. A. Westerberg, *Anal. Chem.*, 1969, **41**, 1770.
48. E. Proksch, H. Bruneder, and V. Granzener, *J. Chromatogr. Sci.*, 1969, **7**, 473.
49. B. G. M. Vandeginste and L. DeGalan, *Anal. Chem.*, 1975, **47**, 2124.
50. J. Novak, K. Petrovic, and S. Wicar, *J. Chromatogr.*, 1971, **55**, 221.
51. J. P. Foley, *J. Chromatogr.*, 1987, **384**, 301.
52. R. J. Hunt, *J. High Res. Chromatogr., Chromatogr. Commun.*, 1985, **8**, 347.
53. J. M. Davis and J. C. Giddings, *Anal. Chem.*, 1983, **55**, 418.
54. M. Martin and G. Guiochon, *Anal. Chem.*, 1985, **57**, 289.
55. A. H. Anderson, T. C. Gibb, and A. B. Littlewood, *Chromatographia*, 1969, **2**, 466.
56. A. H. Anderson, T. C. Gibb, and A. B. Littlewood, *J. Chromatogr. Sci.*, 1970, **8**, 640.
57. J. T. Lundeen and R. S. Juvet, Jr., *Anal. Chem.*, 1981, **53**, 1369.
58. Z. Hippe, A. Bierowska, and T. Pietryga, *Anal. Chim. Acta*, 1980, **122**, 279.
59. B. Vandeginste, R. Essers, T. Bosman, J. Reijnen, and G. Kateman, *Anal. Chem.*, 1985, **57**, 971.
60. M. Goedert and G. Guiochon, *Chromatographia*, 1973, **6**, 39.
61. B. Jover and J. Juhasz, *J. Chromatogr.*, 1978, **154**, 226.
62. D. W. Grant and A. Clarke, *Anal. Chem.*, 1971, **43**, 1951.
63. S. P. Cram, S. N. Chesler, and A. C. Brown, *J. Chromatogr.*, 1976, **126**, 279.
64. R. W. McCoy, R. L. Aitken, R. E. Pauk, E. R. Ziegel, T. Wolf, G. T. Fritz, and D. M. Marmion, *J. Chromatogr. Sci.*, 1984, **22**, 425.
65. A. N. Papas, 'Chromatographic Data Systems: a Critical Review', *CRC Crit. Rev. Anal. Chem.*, 1989, **20** (6), 359.
66. Unpublished data.

# CHAPTER 2

# Manual Measurement of Peaks

## 1 Representation of the Detector Signal by Chart Recorder

Strip chart recorders have been discussed elsewhere.[1,2,3] The basic requirements of a good recorder are a fast linear response time of less than a half second full scale deflection, high input impedance, variable spans including 0–1 mV and 0–10 mV, and chart speeds covering the range 0.5–10 cm min$^{-1}$. It should also have a pen that draws a continuous fine line and preferably one that does not run dry (this does not exist). Commonly used chart widths are 25–30 cm.

When peaks are to be measured manually they are normally recorded at fast chart speeds and at attenuations selected to maximize peak height without overshooting the chart.

Strip chart recorders can distort the detector signal in a number of ways:

**Slow Response Time**
The time taken by the pen to deflect full scale is called the response time. It is determined by the system electronics, and the inertia of the pen head and the drive mechanism.

If the response time is slow, the pen will lag behind the detector signal as a peak emerges and 'meet it on the way down'. If the detector signal returns to baseline as fast as it departed, the pen will lag behind the signal on the way down too (Figure 2.1). The recorded peak will be smaller in height and area than it should be, and the retention time will be increased (Figure 2.1).

Recorders used for chromatography have a response time of 0.5 seconds or less. Peak widths measured at half height must be slower than this.

**Non-Linear Signal Response**
As the pen head draws a peak it moves through periods of acceleration, maximum speed and deceleration which creates a non-linear response over the chart width (Figure 2.2) rather than the linear response suggested by Figure 2.1. Distortion resulting from slow and non-linear response can escape visual detection, especially with WCOT peaks.

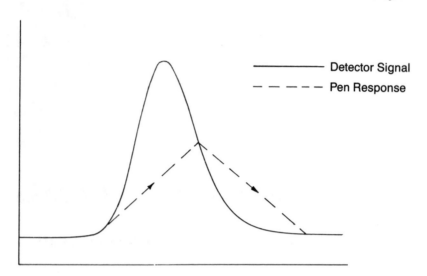

**Figure 2.1**   *When recorder response is slow, the pen lags behind the signal*

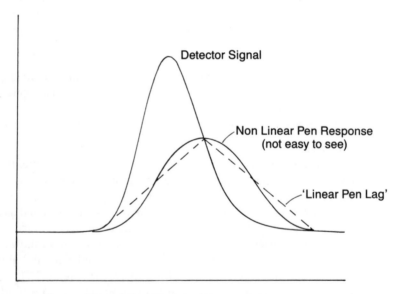

**Figure 2.2**   *Non-linear slow recorder response*

**Pen Head Damping**
Damping circuitry in the form of RC filters is built into the system electronics to prevent pen overshoot if the signal changes suddenly (when changing attenuation or recorder span for example). It adds 'extra column broadening' to fast peaks but does not change peak area (Figure 2.3).

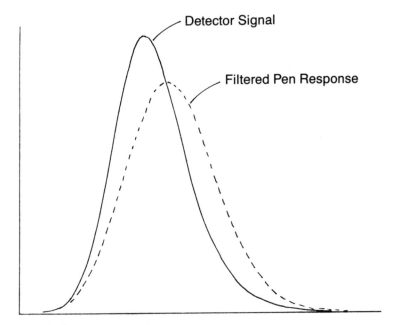

Detector Signal

Filtered Pen Response

**Figure 2.3** *Effect of 'RC' noise filtering. Areas are the same; shapes, heights, and retention are not*

## Amplifier Noise

All electronics have some random noise which adds to the system noise superimposed on the solute signal. Such noise makes it harder to distinguish the start and end of peaks from the background noise. In a well maintained recorder of good specification, this type of noise is only a problem when measuring trace components where the highest sensitivities are required.

So far, strip chart recorders have contained little computer logic to allow software smoothing (though this will change). Noise is filtered electronically (by more RC filters) resulting in further peak broadening and skewing of shape.

## Dead Band

Recorder dead band is the largest change in input signal which does not produce a pen deflection. Laboratory recorders in good working order have relatively small dead bands which can be ignored in practice; worn mechanical parts make them larger. When the baseline drifts off the bottom of the chart and subsequently fails to record baseline events, it is in effect a dead band error.

Dead bands cause the onset of a positive slope to be recorded late, the loss of negative slope at baseline to be recorded early and area to be lost. The minimum detectable quantity is larger than it should be. In effect, a slice has been shaved off the bottom of the chromatogram (Figure 2.4).

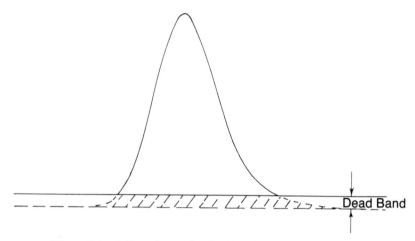

**Figure 2.4**   *Effect of recorder dead band. Shaded area is lost*

Losing 1% from the base of a Gaussian peak through dead band error reduces the peak area by 2.66%.[4] Errors in small peaks are proportionally even greater as dead bands are fixed in size. Peaks may be lost altogether.

### Chart Motor Control

The steady movement of chart paper, essential for time measurement, is maintained either by a discontinuous stepper motor or a continuous synchronous motor. Since it is common practice to improve measurement precision by recording peaks at a fast chart speed, errors of non-uniformity in chart movement directly affect peak area but not peak height. It has been found that precision of width measurement increases with peak width but only to a limiting value; no matter what the height of the peak, maximum width precision will have been achieved once the width at half height reaches 5–6 cm.[5]

Capillary chromatographers have found that the pulsed motion of some stepper motors creates apparent shoulders on fast peaks. When the number of movement pulses per motor revolution is too low, the steps can be seen.

### Attenuator Accuracy

Signal attenuation to maximize peak size requires accurate attenuation. Any errors will be directly incorporated into height or area measurements and both will be affected equally.

The amount of allowed peak expansion is determined by the recorder time constant. Changing the span to increase the height of small fast peaks requires the pen to travel much further in the same time; the time constant may not be fast enough to allow this without distortion.

## 2 Measurement Strategies

There are two approaches to manual measurement of chromatograms: use of methods which assume no peak shape, or use of methods based on a peak model, usually Gaussian.

Counting squares, cutting and weighing, and planimetry are known as 'boundary methods'. These have long been used by engineers, cartographers, *etc.*, and were quickly adopted by chromatographers. Boundary methods make no assumptions about shape and are therefore equally suitable for symmetric and asymmetric peaks. They are, however, tedious to use and are not generally employed in working laboratories.

### Counting Squares

The chromatogram is recorded on 'squared' chart paper with each peak size maximized by attenuation and chart speed. Baselines, perpendiculars and tangents are drawn by the analyst who then counts the squares enclosed within each peak boundary.

The count is made in two stages: first, the number of squares not intersected by a boundary is counted, and to this total is added the sum of part squares through which the peak boundary does intersect.

Counting part squares involves matching those parts which, in the analyst's judgement, add up to a whole square. The matched squares are then added to the sum of the whole squares. Each peak area is multiplied by the attenuation at which it was recorded, and composition of the solution is calculated.

Counting squares requires too much subjective judgement and is completely impractical when there are many peaks. It is not very accurate though practice can bring precision.

### Cutting and Weighing

Peaks are carefully cut out and weighed in an accurate balance. If paper thickness and moisture content are uniform the weight is proportional to the peak area.[6] Each weight is multiplied by the attenuation at which the peak was recorded and the composition of the mixture is then calculated.

Small but important peaks riding on the tails of large peaks are a problem. If the small peak is enlarged for ease of measurement, it will be at the expense of the large peak on which it rides.

Chromatograms can be photocopied and the copy worked upon, thus preserving the original trace. Errors in cutting out peaks are not disastrous because the measurement can be repeated using another copy. If the photocopying paper is a heavier grade than the original chart paper, and its weight and moisture content are equally uniform, it will give larger weights for the peaks and so reduce the weighing error.

Photocopiers with zoom facilities offer a convenient solution to the tailing peak problem, but magnifying the peak size also magnifies the width of the recorder pen line with consequent loss of measurement precision. It is prudent to check the magnification of the photocopy rather than trust the manufacturer's figure, and small peaks should not be magnified repeatedly or optical distortion will come into effect.

**Planimeters**[7]

A planimeter is a mechanical device for measuring the area of an irregular plane shape. Briefly, the shape of the peak is traced out by a stylus linked to a wheeled scale. When the stylus has traced the complete peak perimeter, the area is read from the scale as the difference between the initial and final readings.

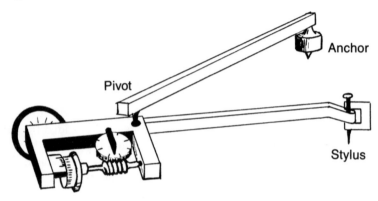

**Figure 2.5**  *Planimeter*

Accuracy is achieved by repeating the measurements and averaging them. Planimeters measure peaks more quickly than counting squares or by cutting and weighing, and can be more precise. The technique is tiring because of the concentration and care required. Good lighting is important and working on a polished surface helps by allowing the user's arm to slide smoothly when tracing the peak.

Errors from the planimeter occur when the measuring wheel skids over the surface of the chromatogram, and this happens, in practice, when the angle between the planimeter arms becomes too small or the user's wrist or elbow slips.

# 3   Measurements Based on a Peak Model

If a peak can be assumed to have a known shape, its area can be calculated from peak dimensions (see Chapter 1). Any departure of the peak shape from the peak model introduces errors of accuracy no matter how precise the measurements. These errors can be offset by calibration, but it is always an implicit assumption of this type of calibration that both measurements will be equally wrong and the errors will cancel. It is better to improve the chromatography so that such strategies are unnecessary.

Most basic theory of peak measurement comes from an assumption of

Gaussian shape, yet few real peaks are Gaussian, many are asymmetric and none has the infinite boundaries which Gaussian theory predicts.

Alternative models have been sought and of these the exponentially modified Gaussian (EMG) function has found the widest applicability so far, though it has its limitations too. It does not fit every peak and has infinite limits.

### Pencil and Rule Methods

Pencil and rule methods measure various peak heights and widths and compute solute quantity from them. They are quicker than boundary methods but are less accurate for overlapping or asymmetric peaks. Measurements based on the exponentially modified Gaussian peak shape give better accuracy for asymmetric peaks once the peaks have been tested for EMG suitability. Pencil and rule methods include:

(1) peak height measurement;
(2) triangulation at peak asymptotes;
(3) peak height × half width, including Condal-Bosch variation[8] and Foley variations (EMG)[9].

### Measurement of Peak Height

Peak height measurement was originally intended to be a simple and fast technique relying on the premise that, for peaks of fixed shape and width, area is proportional to height. This precludes direct peak comparisons within the same chromatogram where width is related to retention time, but peaks can be calibrated by external standards of the same retention and width.

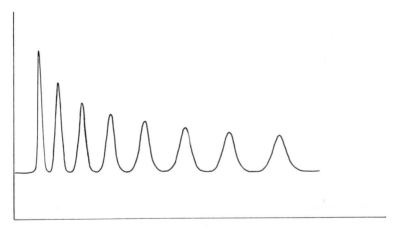

**Figure 2.6** *Peaks of equal area but different heights*

### Height vs Area Measurements

The great disadvantage of manual measurements compared to digital integration are slowness and imprecision. Measurement of height reduces the processing time (and tedium), making it the preferred alternative. If peaks are symmetrical and well resolved there is no sacrifice of accuracy and height is measured with greater precision because fewer constructions and measurements are required and so fewer measurement errors are made.

Noise affects the accuracy of height measurements less than area on chromatograms which are noisy ($S/N < 30$ approximately[10]), but otherwise height and area are equally good.

Peak asymmetry and detector non-linearity perturb height measurements more than area[11] and where these are major factors, peak areas should be measured.

### Triangulation

Lines are drawn through the points of inflection on each side of the peak (points of maximum slope) to intersect and form a triangle with a baseline drawn beneath the peak. The tangents should not intersect at an angle less than 30° (approximately) as this adds to the imprecision of height measurement, nor greater than 120° (approximately) as this leads to imprecision in measuring base width[5] and width at half height.

The area of the triangle calculated from the base width and height is proportional to the peak area (see Chapter 1):

$$\text{Area} = 0.5\,H'w_b/k \qquad k = 0.968 \tag{1}$$

Actual peak height ($H$) can be measured, in which case $k = 0.798$.

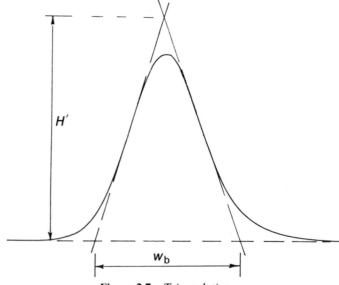

**Figure 2.7**   *Triangulation*

Triangulation is valid for symmetrical peaks on a level baseline (Figure 2.7). Asymmetry and overlap introduce inaccuracies; the worse the peak shape or overlap, the greater the error.

When overlap and asymmetry are not severe, two overlapping peaks can be measured by triangulating the unresolved half of each peak and doubling it to obtain the whole area (Figure 2.8).

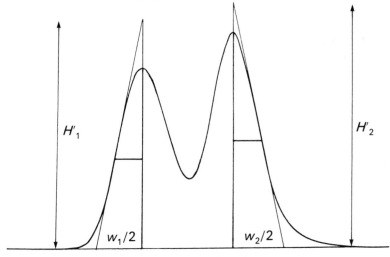

**Figure 2.8** *Triangulation of overlapping peaks*

For asymmetric peaks, doubling the 'half area' does not give an accurate measure of the total area. If the asymmetry of each peak is similar, an alternative strategy is to scale each half area to the height of the other and add it to the other half.

Calibration errors occur when the peak shape changes with injected sample volume (overloading) and the relationship between constructed triangle and peak area is not linear. A multipoint calibration is required.

**Peak Height × Width at Half Height**

Symmetrical peak areas are computed from the product of peak height and the peak width measured at half height (Figure 2.9).

$$\text{Area} = Hw_{0.5}/k \qquad k = 0.939 \qquad (2)$$

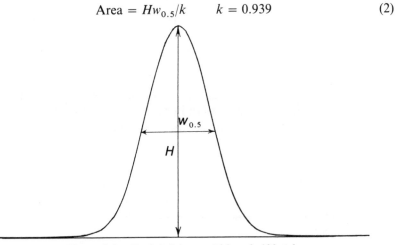

**Figure 2.9** *Peak height × width at half height*

The method works well for resolved and symmetrical peaks but asymmetry and overlap reduce accuracy. The width at half height is easy to locate and measure, but it is not the optimum width for maximum measurement precision.[5] Determination of this optimum width is complicated, and depending on peak shape it will occur at different heights, making it an impractical measure to use, and the width at half height is used instead.

Area measurement of a known peak can be speeded up by attenuating the peak top by half as it elutes (Figure 2.10).

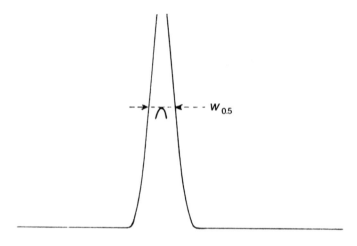

**Figure 2.10**   *Attenuating the peak top by half shows where to measure* $w_{0.5}$
(Reproduced with kind permission from 'Techniques of Organic Chemistry', Vol. XIII, Interscience Pubs., 1968)

A horizontal tangent across the peak maximum provides the width at half height to be measured.[3] Obviously attenuation must be accurate and must not take place until the detector signal has passed halfway to the peak maximum.

Measurement of peak width on a sloping baseline requires some geometrical construction to ensure that the width is always measured horizontally (Figure 2.11).

Measurement of area or height on a sloping baseline is subject to error when the baseline is curved and not linear as constructed. Any attempts to estimate the curvature by eye are likely to be subjective and it is best to use the straight line.

### Condal-Bosch Variation[8]

A simple method designed to compensate errors of asymmetry and measure the actual peak area ($k = 1$) was devised by Condal-Bosch. Instead of using the peak width at half height, the average of the widths at 15% and 85% height is used (Figure 2.12).

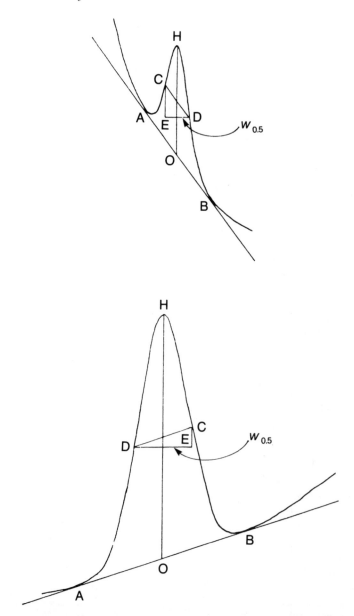

**Figure 2.11**   *Width at half height must be measured horizontally. CD is drawn parallel to AB through $\frac{1}{2}$ HO. ED is horizontal*

$$\text{Area} = H\,0.5[w_{0.15} + w_{0.85}]/k \qquad (3)$$

since $k = 1$,
$$\qquad\qquad = H\,0.5[w_{0.15} + w_{0.85}] \qquad (4)$$

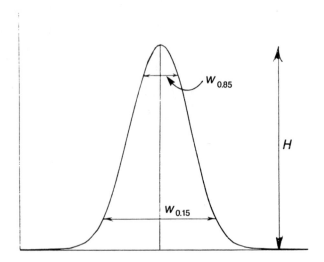

**Figure 2.12**    *Condal-Bosch area measurement*

For any other width combinations of '$x$%' and '$(100 - x)$%, the constant $k$ is not equal to 1 (Figure 2.13).

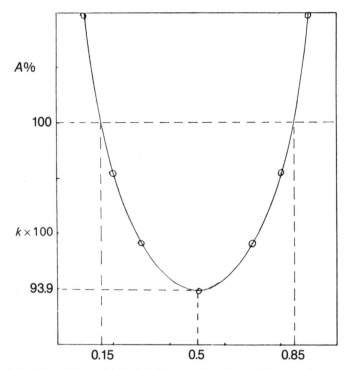

**Figure 2.13**    *The 0.15 and 0.85 heights are the only combination that measure the true peak area (i.e. 100% A). Width at half-height gives 0.939 A (see Equation 2)*

The disadvantage of this method is that it is limited to peaks sufficiently resolved to allow the 15% width to be measured with confidence.

### The Foley Variations[9]
Foley developed equations for peak area based on the EMG function. He showed that for 90% of the asymmetric (LC) peaks in his experiments, peak area could be expressed by:

$$\text{Area} = 0.753 H w_{0.25} \tag{5}$$

where $H$ is the peak height and $w_{0.25}$ is the peak width at 25% of its height (Figure 2.14).

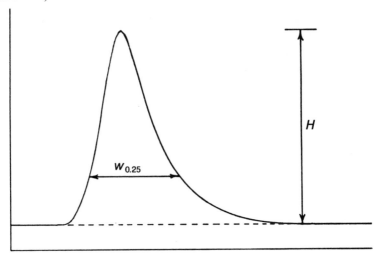

**Figure 2.14**  *Foley (EMG) measurement of peak area*

Equation 5 uses only one width measurement so the calculation is no less convenient than height × width at half height. It is quicker than Condal-Bosch, and the 25% height suffers less from peak overlap than the 15% height.

### Measurement of Overlapping Asymmetric Peaks
Other equations measure the areas of overlapping asymmetric peaks from width measurements taken at greater peak heights to evade encroaching overlap. These require independent measurement of the asymmetry ratio, $B/A$,[9] but they are more accurate than equations which assume Gaussian shape.

$$\text{Area} = 1.07\, H\, w_{0.5}\, (B/A)^{0.235} \tag{6}$$

$$\text{also} \quad = 1.64\, H\, w_{0.75}\, (B/A)^{0.717} \tag{7}$$

### Peak Shape Tests
To prove EMG suitability, shape tests based on area are recommended.

Areas calculated from the three different Equations 5, 6, and 7 should be within 2–3% of each other. Similar tests should also be applied to peaks before using Gaussian equations, of course, but they rarely are.

## 4   Errors of Manual Measurement

Delaney[13] used an EMG function to generate computer peaks of increasing asymmetry ($\tau/\sigma$ ranged from 0 to 4) and studied the determinate errors in using the foregoing equations to measure peak area. The results of this study are summarized in Figure 2.15.

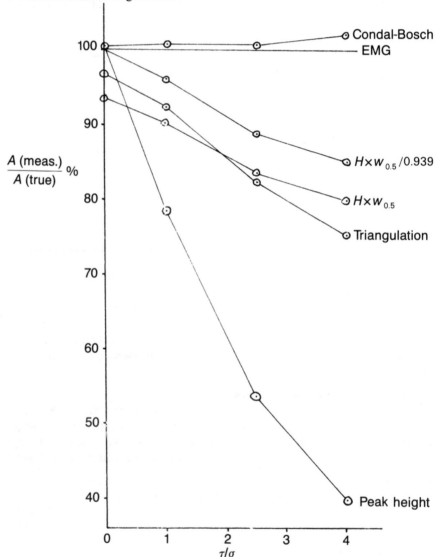

**Figure 2.15**   *Effect of peak asymmetry on manual measurements.*
(Reproduced with kind permission from *Analyst*, 1982, **107**, 606)

Condal-Bosch was accurate to within 2% over the whole range. The EMG equations are accurate because the measured peaks are EMG based. All the Gaussian based methods become increasingly inaccurate, reporting low areas, with height showing the most sensitivity to peak shape. Triangulation starts off being more accurate than height $\times$ width(0.5) but becomes worse at higher asymmetries ($\tau/\sigma > 2$) because the peak base broadens faster than the width at half height.

Figure 2.15 additionally implies that peaks should only be calibrated against others having the same shape, and this in turn implies calibration against peaks of the same size if shape varies with injection volume.

Ball *et al.*[5,14] made a systematic study of the geometrical construction errors made when peaks are measured manually. The principal constructions are shown in Figure 2.16.

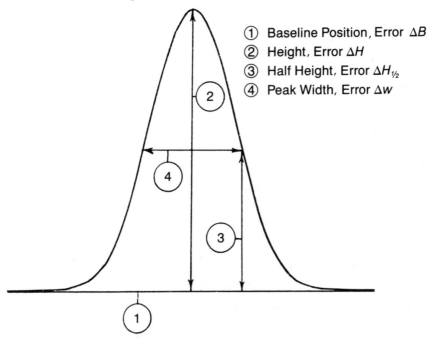

① Baseline Position, Error $\Delta B$
② Height, Error $\Delta H$
③ Half Height, Error $\Delta H_{\frac{1}{2}}$
④ Peak Width, Error $\Delta w$

**Figure 2.16**   *Manual measurement errors*

With each step there is an associated measurement or placement error. The error in measuring peak area is a combination of all these errors. In general the variance of the area measurement is the sum of the component variances and so:

$$\frac{\Delta A}{A} = \sqrt{\left(\frac{\Delta B}{A}\right)^2 + \left(\frac{\Delta H}{A}\right)^2 + \left(\frac{\Delta H_{\frac{1}{2}}}{A}\right)^2 + \left(\frac{\Delta w}{A}\right)^2} \qquad (8)$$

Peak height only contains the errors in positioning the baseline and measuring height:

$$\frac{\Delta H}{H} = \sqrt{\left(\frac{\Delta B}{H}\right)^2 + \left(\frac{\Delta H}{H}\right)^2} \qquad\qquad (9)$$

Comparing equations 8 and 9 confirms that peak height precision will be greater than that of area because fewer measurements and errors are involved.

## Optimum Peak Shape

There is an optimum peak shape for maximum area precision which occurs when the ratio of peak height to width at half height is in the range 2 to 5 (Figure 2.17).

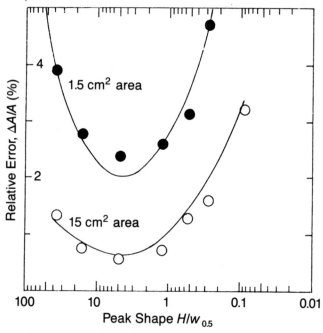

**Figure 2.17**   *Change in precision of peak area with change in ratio of height to width for peaks of different areas*
*(Reproduced with kind permission from Anal. Chem., 1968, **40**, 129)*

**Table 2.1**   *Comparison of Manual Methods*

| *Method* | *Precision%* | *References* |
| --- | --- | --- |
| Cut & Weigh | 1.7 | 15, 16 |
| Planimetry | 0.5–4 | 6, 15, 16 |
| Height | 0.25–0.9 | 5, 16 |
| Triangulation | 1.1–4.1 | 9, 15, 17 |
| Height × w(0.5) | 0.8–2.5 | 5, 15, 16, 17 |
| Condal-Bosch | 2 | 8, 13 |
| EMG | 1.4–2 | 9 |

Precision is greater for large peaks than for small (fractional error is inversely proportional to $\sqrt{A}$) but there is a limit to the increase in precision achieved by increasing chart speed. Precision increases until the width reaches 3–5 cm but then becomes constant, independent of peak size (Figure 2.18).

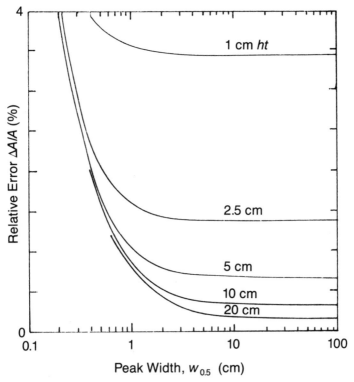

**Figure 2.18** *Variation of precision with peak width for peaks of different heights* (Reproduced with kind permission from *Anal. Chem.*, 1968, **40**, 129)

There is no optimum height. Precision of height measurement increases with peak height. Provided that peaks are symmetrical the fractional error is inversely proportional to peak height, so for a given area, the narrower the peak the better.

A comparison of the precision of methods is given in Table 2.1. Height is the best and triangulation generally the worst.

## Advantages and Disadvantages of Manual Peak Measurement

All manual measurements are less convenient than automatic methods because the chromatographer is required to do some work. However, using a good chart recorder has some real benefits:

(1) The chromatogram is the most direct representation of the detector signal. It has not been digitized, processed and reconstructed as happens in integrators and computers.

(2) The analyst makes the decisions where peaks begin and end, and where baselines, perpendiculars and tangents should be drawn. The choice between perpendicular and tangent separation of marginally sized peaks can be made consistently over a series of experiments.

(3) The analyst is better than a computer in discriminating between noise and peaks, and in observing and rejecting peaks which are not part of the current analysis.

(4) Manual methods are cheap; only the planimeter need be purchased.

(5) Processing of chromatograms can be done at any convenient time, even away from the laboratory.

(6) If peaks do not go 'off scale', the analysis can be left unattended.

There are disadvantages:

(1) Manual measurements require great care to achieve good precision. The methods are time consuming and the results are not as precise as those measured by integrator or computer.

(2) Small peaks on tails which cannot be enlarged are measured with even less precision.

(3) Overlapping peaks are not reliably measured though manual EMG based measurements can be more accurate than integrator measurements which use perpendiculars and tangents.

(4) If important peaks do go 'off scale', the analysis must be repeated which is not always possible if no sample is left, or it has degraded.

(5) Measuring peaks manually is always tedious.

## 5 References

1. R. L. Grob, 'Modern Practice of Gas Chromatography', 2nd Edn., J. Wiley and Sons, New York, 1985.
2. R. G. Bonsall, *J. Gas Chromatogr.*, 1964, **2**, 277.
3. O. E. Schupp, 'Gas Chromatography', 'Techniques of Organic Chemistry', Vol. XIII, Interscience, New York, 1968.
4. A. B. Littlewood, *Z. Anal. Chem.*, 1968, **236**, 39.
5. D. L. Ball, W. E. Harris, and H. W. Hapgood, *Anal. Chem.*, 1968, **40**, 129.
6. W. J. A. Vanden Heuval and A. G. Zacchei, 'Advances in Chromatography', Vol. 14, Marcel Dekker, New York, 1976.
7. J. Janik, *J. Chromatogr.*, 1960, **3**, 308.
8. L. Condal-Bosch, *J. Chem. Educ.*, 1964, **41**, A235.
9. J. Foley, *Anal. Chem.*, 1987, **59**, 1984.
10. T. Petticlerc and G. Guiochon, *J. Chromatogr. Sci.*, 1976, **14**, 531.
11. J. J. Kirkland, W. W. Yau, H. J. Stoklosa, and C. H. Dilks Jr., *J. Chromatogr. Sci.*, 1977, **15**, 303.
12. Personal communication.
13. M. F. Delaney, *Analyst*, 1982, **107**, 606.
14. D. L. Ball, W. E. Harris, and H. W. Hapgood, *Sep. Sci.*, 1967, **2**, 81.
15. H. M. McNair and E. J. Bonelli, 'Basic Gas Chromatography', Varian Associates, Walnut Creek, CA, 1968.
16. J. M. McGill, F. Baumann, and F. Tao, 'Previews and Reviews', Varian Associates, Walnut Creek, CA, 1967.
17. D. W. Grant and A. Clarke, *Anal. Chem.*, 1971, **43**, 1951.

# CHAPTER 3

# Digital Integrators and Peak Measurement

## 1 A Brief History of Integrators

### Strip Chart Recorder Techniques

As well as the manual techniques of peak measurement described in Chapter 2, attachments to recorders were designed to allow automatic measurement. The most important of these techniques was the 'disk integrator' which went out of large scale production in the mid 1970s but which can still be made to order.

The small trace at the side of the chromatogram in Figure 3.1 is the integral of the detector signal. The disk integrator was widely used but had the same range as the recorder pen. It suffered when the peak was attenuated or went off scale. If the chart speed was too slow the integral trace merged on itself though every tenth deflection was slightly bigger to allow quick counting.

### Electromechanical Counters[1,2]

The rise in signal level above a preset (baseline) value was used to drive a counter whose count rate was proportional to the rise. When the signal returned to baseline, *i.e.* to the preset value, the counter tripped and printed the peak area.

If the signal did not return to baseline, the analyst waited until the end of the peak had emerged on the chart recorder and pressed an 'end' button to force the count to be printed. If the signal drifted below the original baseline level the counter tripped at the preset level and lost the peak tail. The analyst forced area printout at valleys to obtain areas of overlapping peaks.

The disadvantages of electromechanical counters always outweighed the benefits.

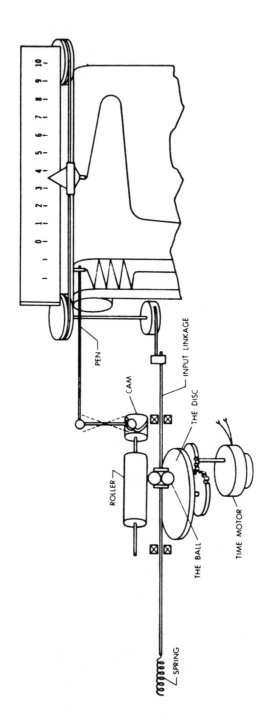

**Figure 3.1**  *Disk integrator*
(Reproduced with kind permission from 'Modern Practice of Gas Chromatography', ed. R. L. Grob, 2nd Edn., Wiley Interscience, 1985)

Disadvantages:

(1) Peaks were only measured above a preset level. This projected a horizontal baseline beneath peaks. The integrator could not cope with drifting baselines (Figure 3.2).

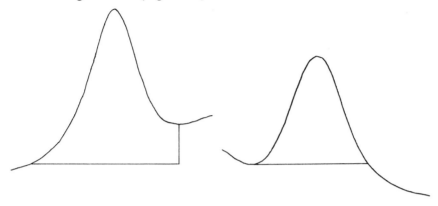

**Figure 3.2** *Effect of sloping baseline on electromechanical integration of peaks*

(2) The analyst was always in attendance to terminate integration at valleys or where the baseline had drifted upwards, and to reset the baseline level just before the start of each peak.
(3) Being electromechanical, the counter had limited dynamic range and a finite count rate which, like a slow time constant, could not keep up with fast peaks and so measured low areas without any warning to the analyst (Figure 3.3).

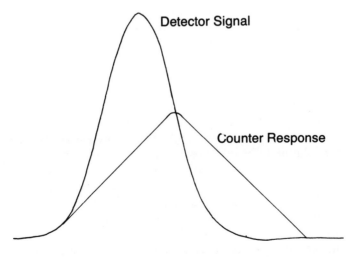

**Figure 3.3** *Counter has finite rate which could lag behind the detector signal*

(4) The baseline zero was adjusted to sit just above the noise level to prevent noise peaks triggering counts. It also sat above the base of peaks which were not included in the count (Figure 3.4).

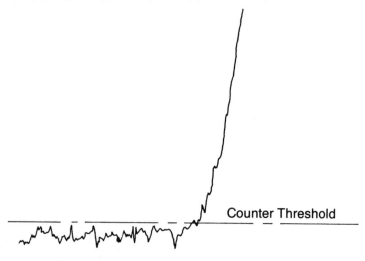

**Figure 3.4**  *Counter zero set just above noise level*

(5) Retention time was not measured: it had to be taken from the chromatogram. The operator added his own notes to the printer report and stapled it to the recorder trace.

(6) There were no calculations. These were later made manually on area counts of limited numeric range.

(7) The integrators were unreliable, and component variations between similar models introduced a repeatability error.

Analysts became distrustful of these instruments and since that time, integrators have indicated with increasing detail how they have measured peaks and from what boundaries.

## Electronic Integrators

### TTL Instruments—the First Electronic Integrators

In 1968 Hewlett Packard launched the 3370 series integrator, the first TTL model with a wide linear dynamic range. This allowed large and small peaks in the same solution to be measured without scale adjustment. It expanded the number of samples which could be processed without intervention by the analyst and could be left to operate unattended with confidence.

The HP 3370 was expensive and incurred competition from similar integrators manufactured by Vidar Autolab (now Spectra Physics) and Infotronics (now Laboratory Data Control).

These integrators were totally electronic. They used voltage to frequency conversion of the analog detector signal but could not store data, which meant that their peak decisions had to be instantaneous. TTL integrators often false-tripped on noise peaks and they could not cope with baseline drift during peak elution. Baseline was simply defined as the last signal level before the start of the peak, and projected horizontally under the peak or group of peaks.

In spite of these limitations TTL integrators were the first 'real integrators' and a large number were sold. Many of their limitations were offset by good chromatography and isothermal analyses (it was mostly GC at that time).

Many modern integrator features were introduced by these instruments:

(1) use of V/F data conversion;
(2) automatic perpendicular separation of unresolved peaks with valley comments printed on the report;
(3) measurement of retention time;
(4) peak detection or 'event' marks to show the start and end of peaks were added to the detector signal and displayed on the chromatogram by the chart recorder;
(5) minimum peak area threshold filtered small peaks;
(6) noise discrimination based on peak width cancelled false peak starts;
(7) slope sensitivity was introduced. The integrator was able to monitor true baseline;
(8) tangent skimming was introduced but not too successfully;
(9) The first calculation, normalization or area% was added;
(10) panel indicators showed experiment status such as reset or analysis, peak integration, positive/negative slope, elapsed analysis time, and peak area.

## Microprocessor Based Integrators[3]

In 1973 Spectra Physics launched the Minigrator, the first microprocessor based integrator. It was followed by the Infotronics 309 series and Varian 100 series integrators.

The Minigrator had memory to store four peaks and place a trapezoidally corrected baseline beneath them. When a larger group was measured, a horizontal baseline was projected under the first few peaks and a trapezoidal baseline drawn under the last four. It monitored trends and updated its parameters to allow for increasing peak width as the analysis progressed.

Tangent skimming of up to three rider peaks was possible (the solvent peak made up the fourth), and an inbuilt calculator allowed the analyst to make a variety of manual calculations such as area% and internal standard.

The integration report was obtained from a printer, but the chromatogram was still displayed on a separate chart recorder.

## Integrators with Printer Plotters

Barely two years after the launch of the Minigrator, Hewlett Packard launched the 3380 series integrator with its own inbuilt chart recorder. The chart recorder was in fact an alphanumeric printer plotter which presented results and chromatogram on one piece of paper.

Peak memory capacity was increased to 54 peaks and calculations of area%, internal standard and external standard included response factors determined by self calibration.

Chart paper advancement stopped at the end of the analysis and saved a lot of paper.

### Modern Integrators and Microcomputers

The Minigrator and the HP 3380 integrators introduced most of the features which exist in contemporary integrators. Developments since then have been almost exclusively electronic. The peak processing algorithms used by integrators have changed in detail but not much in principle; overlapping peaks are still separated by perpendiculars and tangents.

Improvements in peak deconvolution techniques have stalled due to the lack of data contained in an FID or UV detector.[4] The main chromatographic improvements for data processors have come from WCOT technology, by which the analyst steers clear of deconvolution and asymmetry problems rather than have the integrator 'solve' them mathematically.

## The Impact of the Microcomputer

The introduction of microcomputers in the late 1970s and in particular the arrival of the 16 bit IBM PC in 1982 has given rise to an alternative type of software based integrator.

Integrators are dedicated data processors; computers are inherently more flexible, they change their role with their software. The initial attraction of a computer based integrator over a dedicated one is based on two features:

(1) Most scientists want a microcomputer. It is an instrument of clear technological importance which needs to be mastered. Individual purchases within large companies have sometimes been restricted because of concern over uncontrolled choice and the lack of systems analysis skills, even though the hardware costs are comparable to or less than those of an integrator. Integration software is a legitimate way to justify the purchase.

(2) Computers have many other uses made possible by a wide range of available software.

In spite of this, integrators continue to dominate the market though their share is declining. This slower than expected development reflects the number of programmers engaged in writing integration software, rather than the number of computer manufacturers or cost of hardware. In this comparison, integrator designers have a considerable lead over the programmers. Integrator programs have been developed for some time. They are comprehensive, bug free and easy to use.

Computer software is catching up to the standards set by integrators but much of the early software was disappointing and after an initial surge of sales the pendulum swung back to dedicated integrators. It is expected to swing again towards computers as they become information systems in which integration is only a part.

The effect of microcomputers has been to reduce the cost of integration and to shape stand alone integrators to images of computers. Disk storage and screen manipulation of chromatograms are now features of many integrators; BASIC programming, originally added to integrators to allow analysts to make non-standard calculations, adds to the computer image. When analysts need additional power, integrators can interface to a computer to give the benefits of both instruments.

As prices come down, it is likely that integrators will be built into the chromatograph thereby saving the costs of a separate power supply, instrument case, keyboard, *etc*. When this happens, the role of the microcomputer will change towards a general, though small, LIM system retaining the flexibility to perform specific tasks such as integration and report distribution while allowing the analyst to append novel software.

## 2 Current Integrator Status

Any description of a 'modern integrator' will quickly outdate itself, but it is still worth assessing the current state of development in order to review what has been achieved and speculate on what is yet to come.

Integrators now come in four guises:

(1) integrators incorporated into a chromatograph;
(2) stand alone integrators;
(3) microcomputers with integration software;
(4) as components of larger laboratory computer systems.

The first three are simple variants with the same integration role and are directly comparable. LIM systems are rarely bought for their integrator specifications, but any laboratory paying the price will expect a performance at least equivalent to a much less expensive integrator.

### The Standard Integrator Specification

This includes:

(1) measurement of peak areas, heights and retention times with method files to store data processing parameters;
(2) standard analysis calculations of area% (normalization), internal and external standards methods, and calibration or automatic determination of response factors from standard mixtures;
(3) presentation of the chromatogram and analysis report to an alphanumeric printer plotter;
(4) storage of chromatograms for reprocessing.

All current production model integrators offer these facilities.

The influence of micro-computer technology on integrators has additionally added computer facilities such as:

(1) BASIC programming—for non-standard calculations and different report formats.
(2) 'Intelligent communications ports' which link to external computers (usually via BASIC).
(3) Computer peripherals:
 (a) VDUs provide an alternative to the printer for viewing the chromatogram;
 (b) permanent disk storage and disk operating systems (DOS).
(4) The role of system controller. Where the chromatograph and integrator are supplied by the same manufacturer, a more comprehensive communication link can allow the integrator to control and monitor the chromatograph. Indeed the integrator may control more than one chromatograph simultaneously.
 An integrated system can monitor system readiness and column performance, and assess system suitability and results validation.
(5) The most direct influence of microcomputers is of course their use, with appropriate sofware, as integrators.

## Integrator Files

Integrators equipped only with RAM storage typically hold ten files. Each file is able to store all of the parameters and information for one analysis. When a file is selected it immediately configures the integrator to process the chromatogram derived from that analysis. With disk storage the number of stored files is limited only by the disk capacity.

Files hold four types of parameters for:

(1) measurement of peaks;
(2) instrument control and communications;
(3) calculations, calibrations and solute identity;
(4) report preparation.

## Analysis Parameters for Peak Measurement

Parameters which affect measurement of areas, heights, retention times, *etc.*, are discussed in detail later. They include the detector signal sampling frequency or 'peak width' parameter, 'slope sensitivity' and 'baseline drift tolerance' parameters.

Analysis parameters are normally optimized during method development and only changed if necessary between analyses. They can be adjusted during an analysis by deploying a Time Program which is a set of timed commands to change parameter values at prescribed times after injection. It rids the analyst of the need to attend the analysis in order to make the changes himself. At the end of the analysis the parameters return to their initial values.

Before storage and reprocessing of chromatograms became available, Time Programs included commands to start the autosampler, operate relays, switch valves, change columns, *etc*. These are now stored in a separate Event or Analysis Control Program which is only employed during real time analysis, and not during data reprocessing. Stored chromatograms can be reprocessed without re-enacting the experiment.

The role of analysis parameters is to optimize measurement of required quantities such as peak area, and enable the integrator to ignore unwanted events such as baseline disturbances. All real peaks are reported, but not the injection pressure pulse on the solvent peak, or the reagents added during derivatization.

Analysis parameters are also used to over-ride the integrator's inclination to ruin peak measurement by establishing baseline in wrong places, for example, at the bottom of a negative baseline excursion.

The most important analysis parameters are discussed below:

### (a) Peak Width
Whatever the name given to this parameter by any individual manufacturer, it is the detector signal sampling frequency, the rate at which the detector signal is digitized for integration or storage. It is the most important integrator parameter.

For manufacturing ease, the electronic sampling frequency is fixed at a high value typically in the region of 100 to 1000 samples per second. It can be made to appear variable by bunching these high frequency samples into a single sample whose value is the same as would have been obtained at a slower sampling frequency. The bunching is software controlled and therefore, in contrast to a variable electronic sampling frequency, adds nothing to manufacturing costs.

The degree of bunching is related to the peaks being measured so that the same detection and measurement algorithms can be used throughout the chromatogram.

The analyst programs the Peak Width parameter to measure early peaks. Later (and broader) peaks are measured by an updated value from the Time to Double function (see below), or by a preset command in the Time Program.

### (b) Slope Sensitivity
Peaks are detected because the detector signal amplitude changes more rapidly when peaks elute than during baseline signal between peaks. Integrators track the detector signal and detect the change in slope which occurs when a peak emerges, but smaller changes are allowed to pass as baseline fluctuations. There is a threshold of slope below which peaks are not detected, and this value is set by the 'slope sensitivity' parameter. Baseline whose drift does not exceed this threshold is considered to have zero slope.

It is usual for the integrators to perform a test (a 'slope' or 'noise' test) to measure and self-program an appropriate value for slope sensitivity.

#### (c) Baseline Drift Tolerance

Accurate peak sensing is only one half of the problem for an integrator, locating baseline at the end of a peak is the other.

Baseline only exists at certain allowed locations, where the analyst would expect it to be. The range of these locations is determined by the 'baseline drift tolerance' parameter. The name varies from one manufacturer to another but the principle is the same. Baseline is defined as zero slope (inside the limits allowed by slope sensitivity) within a range allowed by the drift tolerance. Outside this range, zero slope is not recognized as baseline no matter how long it lasts.

The purpose of this parameter is to allow the integrator to find new baseline after a peak, or group of peaks, has eluted, when the detector signal does not return to its original level. It prevents the integrator fixing baseline at other points of zero slope like peak tops, or in valleys.

Baseline drift tolerance is often misunderstood and confused with slope sensitivity, yet the difference is clear: drift parameter determines how *far* the baseline can drift from its original position, slope sensitivity determines how *fast* it may drift away.

#### (d) Time to Double

Peaks vary in width and shape. Values of peak width and slope sensitivity which are correct for the start of an analysis may not be appropriate for peaks at the end. The 'time to double' parameter allows progressive updating of peak width and slope sensitivity at regular time intervals during the analysis and keeps each near to its optimum value for the emerging peaks.

Time to double is a useful relic from the days of less sophisticated integrators and is a precursor of the time program since its function is to change parameter values during an analysis. Peak width and slope sensitivity could also be changed by a time program, but if these are the only parameters involved, it is easier to set one parameter than make up a time program.

When integrators were first employed, isothermal GC analyses were prevalent and in these analyses, peak widths and retentions of homologues increase exponentially. At an appropriate time interval, the peak width parameter doubles in value and the slope sensitivity halves (*i.e.* becomes twice as sensitive). This doubling up emulates the exponential increase in peak widths and enables the integrator to track better the slower onset of broad peaks which elute late in the analysis.

In the present generation of integrators, time to double is an automatic technique. The integrator monitors peak widths during an analysis, and if the current parameter value is too small, it is increased (though not necessarily doubled) to a better value which matches eluting peaks.

Peak width and slope sensitivity parameters are interrelated. Doubling of peak width implicitly means doubling of the sensitivity: one is not changed without the other. (See p. 129.)

### (e) Minimum Area or Height Threshold

Long term noise and genuine but unwanted small peaks are unavoidably measured but can be removed from the final report by setting a minimum size criterion. This size test is only applied when peak measurement is complete, which is too late to prevent peak detection marks and retention times being added to the chromatogram.

The same threshold applies to height or area, though a filter value appropriate to areas will be about 100 times too large when applied to the heights of the same peaks.

### (f) Analysis Duration

The analysis time is programmable so that the instrument will initiate an analysis report and reset the instrument for the next injection while unattended.

## Parameters which Override the Integrator's Logic

These parameters are mostly used in a Time Program to prevent measurement of unwanted peaks and incorrect placing of baselines.

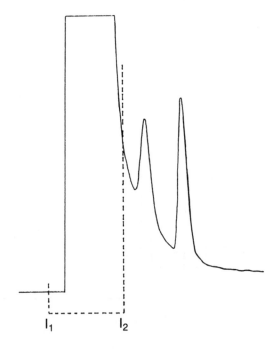

**Figure 3.5**   *Integrate inhibit prevents measurement of solvent peak whose tail is baseline for rider peaks. The signal at $I_2$ is fixed as baseline*

### (g) Integrate Inhibit

Integrate inhibit disables peak recognition for a specified period. It is used to avoid measurement of unwanted solvent peaks, or false triggering of integration by baseline disturbances.

The first signal the integrator sees after a period of integrate inhibit is defined as baseline, whatever is happening to the detector signal.

### (h) Forcing Tangent or Perpendicular Separation

Integrators may separate two fused peaks by a perpendicular when the analyst would prefer tangent separation, or vice versa. This problem arises when the second peak is at that marginal size where small changes in concentration cause it to be separated from its neighbour by a perpendicular in one analysis but by a tangent in the next. This ruins any chance of comparing analyses. By binding the integrator to one type of separation the analyst can at least achieve consistency.

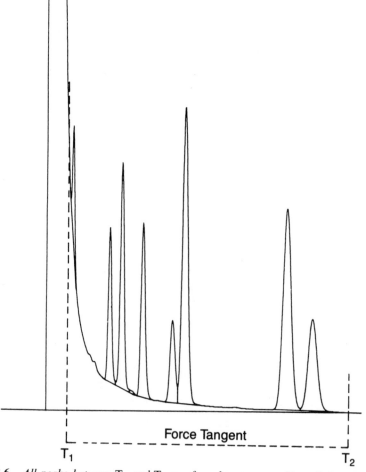

**Figure 3.6** *All peaks between* $T_1$ *and* $T_2$ *are forced to tangent skim off the solvent peak*

## (i) Forcing Baseline Detection

Integrators can be forced to recognize specific baseline points, implicitly to terminate a previous peak detection, or they can be forced to complete peak measurement at a moment determined by the analyst.

The integrator is typically directed to establish baseline after some disturbance has ended, and just before a peak of interest is due to elute.

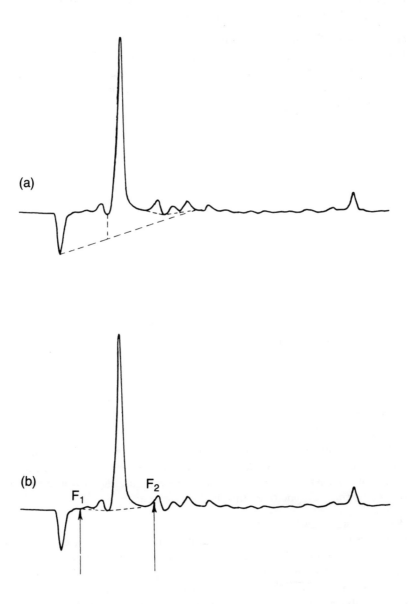

**Figure 3.7** (a) *Without forcing baseline;* (b) *baseline is forced at* $F_1$ *and* $F_2$

Forcing baseline recognition is used to rescue a peak from a poor environment which makes it difficult for the integrator to measure it correctly. The correct analytical approach to solve the problem would be 'improve the chromatography', but even if this can be done it does not rescue a peak in a completed analysis. The analyst selects the best baseline points to isolate the peak from nearby disturbances, makes up a Time Program and reprocesses the data. On many occasions, forcing baseline and integrate inhibit have the same purpose.

### (j) Horizontal Baseline Projection

Baseline disturbances, especially negative excursions, which spoil location of the true boundaries of a nearby peak may be ignored by projecting a horizontal baseline forwards or backwards from a neighbouring stretch of good baseline.

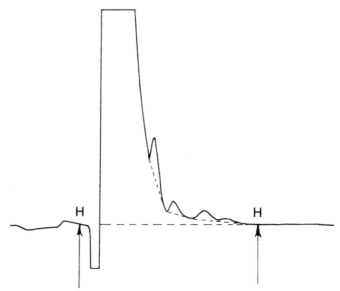

**Figure 3.8**  *Using horizontal baseline does not mean that the solvent peak is measured accurately*

This does not necessarily lead to accurate peak area or height measurement. The analyst merely recognizes that the integrator would produce an even worse measure if left to itself. Like forced baseline, it is a technique to overcome the deficiencies of the chromatography.

### (k) Inverting Negative Peaks

It is possible for a thermal conductivity detector to produce positive and negative peaks in the same chromatogram. Integrators will not normally measure negative peaks because positive slope is required for peak recognition. What is worse, the integrator will track a negative peak to its lowest level, establish baseline there and trigger peak detection as the signal returns to baseline.

Negative peak inversion allows measurement of negative peaks provided they do not breach the lower operation limit of the integrator, about $-5\,\text{mV}$ to $-10\,\text{mV}$, below which they will be clipped.

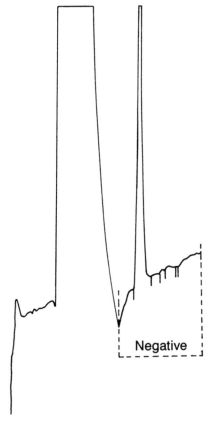

**Figure 3.9**  *Negative peak inversion*

### (l) Instrument Control and Synchronization
Some analyses involve mechanical operations such as autosampling, valve switching or column switching, and require a control or event program to operate the events at the correct times. Relays are a commonly used interface between system modules; they are an uncomplicated technology and reliable.

A more sophisticated form of 'computer' communication is possible when the whole system is from one manufacturer. Monitoring and feedback mechanisms can allow the integrator to control the chromatograph rather than merely synchronize the operations.

## Calculations, Calibrations, and Solute Identity

A third set of parameters, called a 'sample table' or 'solute identity file', is concerned with matching measured peak areas or heights to solute names and assigning response factors or standard concentrations to them.

The parameter table holds data on:

(1) analysis and solute names;
(2) retention times and windows (tolerances);
(3) response factors and standard concentrations;
(4) the type of calculation, area%, internal standard, external standard;
(5) internal standard weights and total sample weights.

For a series of unattended analyses with auto-injection of solutes, this becomes a large file and it is essential that the samples are racked up in the correct analysis order. Bar code readers have been used to prevent mistakes of this kind. They read the sample name and match it to its logged data.

**Peak and Solute Identity**

Integrators identify peaks by their retention time. This identity will have been established by other means, but once the solute is known it is routinely identified in subsequent analyses as the peak which elutes at the expected retention time.

The integrator is a computer, and if instructed to find a peak at 5 minutes, it will look for a peak at 5 minutes but ignore the one at 4.999 or 5.001 minutes. Peaks are therefore identified at a specified time give or take a tolerance.

Tolerance can be defined in two ways: as a percentage or as an absolute time. A peak may be predicted to elute at 5 minutes $\pm$ 10%, which means that any (or all) peaks eluting between 270 and 330 s will be identified as the expected peak. Alternatively, a peak expected at 5 min $\pm$ 0.5 min may elute at any time between 4.5 and 5.5 min and still be identified.

To prevent more than one peak eluting inside the same window, the tolerance is made as narrow as possible consistent with the variation in retention time from one analysis to the next.

**Relative Retention Times**

If absolute retention times are not stable, relative retention times are used. One peak is assigned the role of standard, and every other peak retention is compared to it.

$$\text{Relative Retention} = \frac{t_R}{t_{std.}} \qquad (1)$$

Any uniform retention drift will not affect the retention ratio.

Relative retention times are more stable than absolute retention times. They are less vulnerable to changing column conditions, septum leaks, and similar systematic errors.

The chosen standard must be well separated from its neighbours so that it can be given a wide window from which it is unable to drift. If it does move out of its window it will not be identified correctly, and the whole method breaks down.

## Standard Calculations

Integrators offer three standard calculations; normalization (area%), internal standard and external standard, and a variety of non-standard calculations, some of which use BASIC programs. These are well documented elsewhere[5-9] and will only be summarized here.

## Area% or Normalization

The concentration, $c_i$, of a solute, i, is calculated from:

$$c_i = \frac{A_i}{\sum(A_i)} \times 100\% \tag{2}$$

where $A_i$ is the peak area of solute i, and $\sum(A_i)$ is the total area of all the chromatogram peaks.

If response factors of each solute are included, Equation 2 becomes:

$$c_i = \frac{A_i R_i}{\sum(A_i R_i)} \times 100\% \tag{3}$$

where $R_i$ = absolute solute response factor, the quantity of matter to produce unit area count.

Area% as defined by Equation 3 means percentage of solutes which are eluted and detected. This might only be a fraction of what was injected. The analyst must be certain that nothing has been retained by the column or missed by the detector.

If solutes are known to be retained or not detected, and the measured fraction is known accurately, then instead of normalizing to 100%, the results can be scaled to the fractional percentage by means of a scaling factor:

$$c_i = \frac{A_i R_i}{\sum(A_i R_i)} \times S\% \tag{4}$$

For example, beer is typically analysed by GC using an FID which does not respond to water. If the beer is known to contain 94% water then it is more useful to normalize the analysis to 6% than 100%

## Internal Standard

The weight of solute in a sample is determined by comparing its area to the area of a known amount of standard added to the sample and analysed with it:

$$W_i = \frac{A_i R_{is}}{A_s} . W_s \tag{5}$$

where $W_i$ = solute quantity
$\quad\quad W_s$ = quantity of internal standard
$\quad\quad R_{is}$ = the relative response factor of solute to internal standard
$\quad\quad\quad$ = ratio of absolute response factors

The concentration of solute in the original sample ($W_{spl}$) is given by:

$$\text{wt\%} = \frac{W_i}{W_{spl}}.\,100 = \frac{A_i\,R_{is}\,W_s}{A_s W_{spl}} \times 100\% \tag{6}$$

If some solutes are retained by the column, or not detected, the sum of the detected solute weights will be less than the sample weight, or their concentrations will add to less than 100%.

Internal Standard method allows absolute quantities of solutes to be measured by compensating for variations in injected sample volume. It does not compensate for differences in peak shape between solute and standard if this indicates detector overloading. A good standard gives a peak that is the same size and shape as the peak it is to be compared with. It is a weakness of the internal standard technique that one peak serves as standard for several others where these vary in shape and size.

### External Standard

The external standard method compares a solute peak area in an unknown sample with the area produced by a known amount of solute in a standard sample analysed under identical conditions.
Then:

$$W_i = \frac{A_i}{A_s}.\,W_s \tag{7}$$

No response factor is necessary because the sample and standard are the same species. If the peaks are similar in size, it will avoid peak shape/linearity problems. Equal sized injections of standard and unknown are essential. Manual injections are not normally accurate enough; sample valves or autosamplers are to be preferred.

Standard solutions are normally prepared containing a known amount of each of the solutes required from the unknown solution.

### Calibration

Self determination of response factors from the analysis of prepared standards is featured by all integrators, though the maximum number of standards which can be included in each calibration varies as does the number of repeat injections which are allowed.

The simplest calibrations involve analysis of one or two standard solutions (Figure 3.10).

When three or more standards are analysed there is the choice between multilinear calibration where the graph points are joined up by straight lines, or linear regression (least mean squares) which draws the best straight line through them. At this stage, most manufacturers think it is not worthwhile to offer polynomial curve fitting of data, but it would be an easy addition.

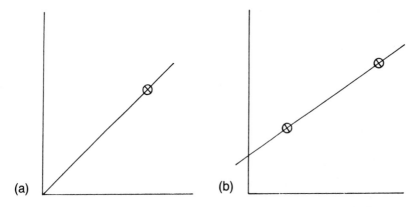

**Figure 3.10** *Calibration. (a) Single standard, y = mx; (b) dual standard, y = mx + c*

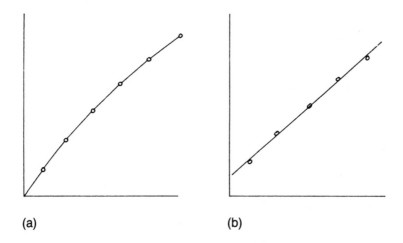

**Figure 3.11** (a) *Multilinear calibration joins the 'dots'; (b) linear regression gives best straight line*

The more standards, the better the calibration, but the more work is involved in preparing them. Single point calibration can give rise to severe calibration errors and should not be used unless the analyst has performed multi-standard calibrations which demonstrated that the calibration is linear and passes through zero.

## Formatting the Analysis Report

The last set of parameters selects from the information generated by the analysis, what is to be printed in the analysis report. The available information includes:

(1) Analysis Notes;
   Date and time
   Operator's name
   Sample or batch name
   Analysis conditions
   Instrument settings
   Sample preparation notes

(2) The chromatogram including peak detection marks and retention times;

(3) Analysis Results;
   Peak areas, heights and retention times
   Peak measurement diagnostics and error flags
   Solute names and concentrations or quantities

(4) Calibration Data;
   The results of calibration experiments, *i.e.* updated sample files
   Response factors

(5) Analysis Validation;
   Column efficiency data
   Peak asymmetry measurements
   'Out of Spec' warnings
   Analysis Abort error messages

## Report Distribution

The analysis report need not include all of the information listed above, but growing legislation requires this kind of information to be available.

When an integrator or computer is linked to a corporate computer it opens up the possibility of distributing the analysis information to several destinations.

(1) The Analyst. The chromatogram and results will be presented for validation in the normal way.

(2) Laboratory Supervisor. Confirms that the work is being carried out and that analyses are within specification, or gives early warning on which samples are failing to match a required specification.

(3) Production Manager. The production manager only wants to know when product of low quality is being made and when corrective action is required.

(4) Accountancy/Personnel. Performance measurement of laboratory staff and instruments allows both to be used optimally.

(5) Permanent Storage. The full set of results is archived to a computer or file server for reference in case of future technical problems when the product history will be reviewed.

# 3   Digital Measurement of Peak Areas

The measurement of chromatographic peaks by computers or integrators is fundamentally different from the manual techniques described in Chapter 2. It is more closely related to the measurement of peak moments in which no assumption of peak shape is made.

Manual measurement of strip chart recorder peaks is made on the analog representation of an analog detector signal by an analyst who makes the logical decisions. Computers and integrators cannot begin to work with the detector signal until it is converted into a digital format. In this conversion, original data is lost, but if the sampling frequency is high enough, the lost information is not critical to the analysis.

## Analog to Digital Conversion[10–12]

The two principal methods of A/D conversion used by chromatography data processors are described below.

### (a) Voltage to Frequency Conversion

Voltage to frequency converters (VFCs) were specifically developed to measure small analog signals such as the millivolt outputs of GC and LC detectors. Although relatively slow, VFCs have been widely used in single channel integrators. Their principal advantages are:

(1) they were designed to measure small voltages (and currents);
(2) wide linear dynamic range;
(3) low drift;
(4) high resolution.

The general schematic diagram for a V/F converter is shown in Figure 3.12:

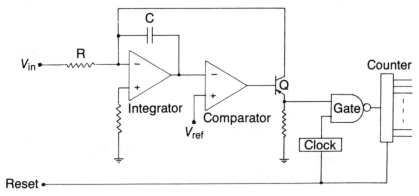

**Figure 3.12**   *Voltage to frequency converter*

The detector signal, $V_{in}$, charges condenser, C, at a rate proportional to $V_{in}$, until the output of the integrator operational amplifier exceeds $V_{ref}$ when the comparator inverts and switches on the transistor Q. This discharges the condenser creating a voltage spike at the emitter of the transistor. The discharge allows the comparator to invert back, switch off Q, and the condenser begins to charge again.

Each charge/discharge cycle generates a pulse at the emitter, and the chain of pulses is gated into a counter. The pulse count within a time window is proportional to $V_{in}$; by controlling the window aperture time, the integrator sensitivity can be varied.

For a typical VFC[13] working with better than 0.005% linearity, the output frequency, $F_{out}$, is related to $V_{in}$ by:

$$F_{out} = 10^5 V_{in} \qquad (8)$$

On baseline the pulse counts are low because the detector signal is low and generally constant (if the baseline is flat). When peaks elute, $V_{in}$ increases and the pulse count increases correspondingly. When the peak has eluted, the signal returns to baseline and the pulse count falls back to the 'baseline rate'.

The pulse count during the window aperture is the integral of the signal voltage over that time and averages out any signal fluctuations occurring within the window.

### (b) Dual Slope, Integrating A/D Conversion

Dual slope A/D converters (Figure 3.13) can sample very rapidly and have been used in conjunction with larger computers to process the signals from several chromatographs simultaneously. As column technology improves and peaks become narrower, they are gradually replacing voltage to frequency converters in single channel integrators.

Dual slope conversion is a two stage process (Figure 3.14).

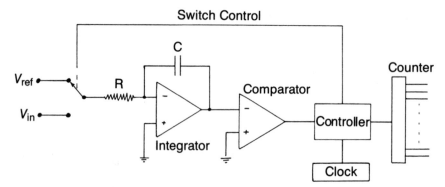

**Figure 3.13**   *Dual slope A/D converter*

In the first stage an integrating operational amplifier is reset and then the detector signal, $V_{in}$, is applied to its input for a fixed time, $t_{in}$. This charges condenser C at a rate proportional to $V_{in}$.

In the second stage, the detector signal is replaced by a reference voltage, $V_{ref}$, which is opposite in polarity to the detector signal. This 'charges down' the condenser at a fixed rate, proportional to $V_{ref}$, and the time required to discharge it is clocked by a stream of pulses into a counter until the discharge of C is complete.

At the end of the count, the integrator and clock are reset, the input is switched back to $V_{in}$, and the time/pulse count proportional to $V_{in}$ is stored (Figure 3.14).

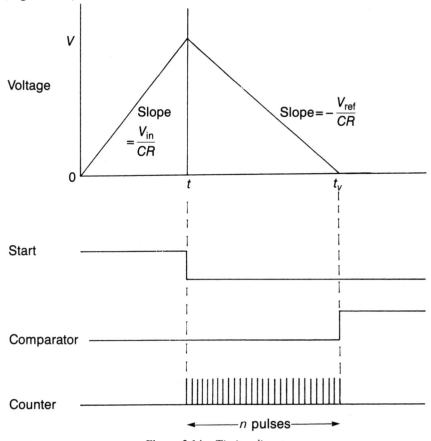

**Figure 3.14** *Timing diagram*

The charge acquired by condenser C from $V_{in}$ is discharged by $V_{ref}$;

$$\therefore \frac{1}{CR} \int_0^{t_{in}} V_{in} \, dt = \frac{-1}{CR} \int_{t_{in}}^{t_v} V_{ref} \, dt \tag{9}$$

If $V_{in}$ and $V_{ref}$ are regarded as the average values of the input and reference voltages during their period of application, then Equation 9 simplifies to:

$$\frac{V_{in}}{CR} \int_0^{t_{in}} dt = \frac{-V_{ref}}{CR} \int_{t_{in}}^{t_v} dt \tag{10}$$

The discharge time is represented by the number of pulses, $n$, from the clock to the counter, *i.e.* $t = nF$, where $F$ = clock frequency;

from Equation 10,
$$V_{in} = \frac{(t_v - t_{in})}{t_{in}} V_{ref}$$

or
$$V_{in} = \frac{n_t}{n_{t_{in}}} V_{ref} \qquad (11)$$

Equation 11 is independent of values for $R$ and $C$. It requires only that $V_{ref}$, $t_{in}$ and the clock frequency, $F$, are constant.

### Resolution of A/D Converters

Resolution is the smallest change in the analog signal which can be seen in the digital output, *i.e.* which changes the least significant bit.

A/D converter resolution is specified by the number of 'bits', *e.g.* 8 bit or 12 bit; a 12 bit A/D converter can resolve to 1 part in $2^{12}$ or 1 part in 4096. If the output voltage is 1 volt, the smallest change in output which the A/D can see is:

$$\frac{1\,V}{4096} = 0.24\,mV$$

Bearing in mind that strip chart recorders are often used at a span of 1 mV, and can show baseline blips of 1% FSD (10 μV), 0.24 mV, or 24% full scale deflection, is not very impressive.

### Auto Ranging of A/D Converters

Commercial integrators and computers use auto-ranging A/D converters which divide a large detector signal into ranges, usually two or three, in order to provide the necessary resolution near baseline where it is needed. For example, with a 12 bit converter, the range 0–10 mV is resolved to 2.4 μV, the 10–100 mV range is resolved to 24 μV and the 100–1000 mV range to 0.24 mV. The necessary calculations to cover range switching are taken care of by other circuitry or software.

## Data Sampling Frequency

Many studies[14–19] have been made to determine the best sampling frequency for measuring peak areas accurately. It is invariably shown that:

(1) measurement accuracy increases with sampling frequency;
(2) errors in asymmetric peaks are greater than those of symmetric peaks sampled at the same frequency.

Theoretically there should be no 'optimum' data sampling frequency. Mathematics predicts and experiment confirms that the faster the data sampling, the more accurate is peak measurement (Figure 3.15). If the sampling rate is too slow, peak detail will be lost and measurement accuracy reduced.

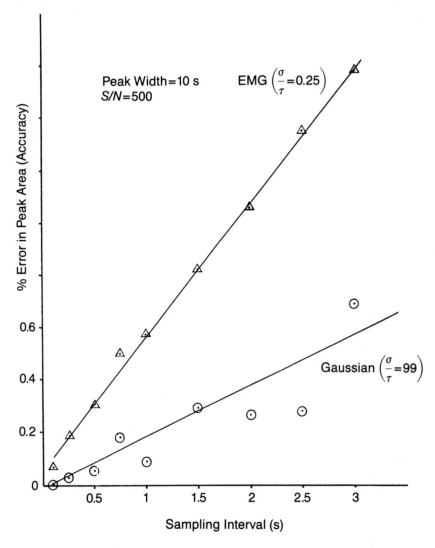

**Figure 3.15** *Accuracy of area measurement improves with sampling frequency* (Data from *J. Chromatogr. Sci.*, 1988, **26**, 101)

There is a practical limitation, however: if the sampling frequency is too high, unwanted baseline noise peaks will be detected and measured. So the sampling frequency is set slow enough to act as a filter for baseline noise. Nyquist sampling theory[20,21] shows that the maximum sampling frequency should not sample the average noise peak more than twice if the noise is to be filtered. The maximum usable data sampling frequency is therefore determined by the baseline noise.

### Sampling Frequency and Integrator Manufacture
It is simpler and cheaper to manufacture an electronic circuit of fixed sampling frequency than one whose frequency is variable. This frequency must be high enough to sample the fastest peaks, but such a rate will generate more data than necessary to process broad peaks, it will require more computer memory to store the extra data, more time to complete data processing, and it will allow detection of more unwanted noise peaks.

### Sampling Frequency and Data Processing Algorithms
Peaks vary in width, generally increasing with retention. If the integrator sampling frequency is fixed, algorithms created to process the stored data must contend with increasing numbers of data samples/peaks.

Simpler processing algorithms can be used if the number of data samples/ peak, or sampling density, is uniform,[22] but this is at odds with the requirement of a simple-to-manufacture, fixed sampling frequency.


## Data Bunching

The conflict between fixed sampling frequency and uniform sampling density is resolved by 'data bunching', a software technique (with zero manufacturing cost) which collects consecutive data samples taken at high frequency into groups or bunches. The number of data samples in the bunch is determined by local peak width,[23,24] and this creates a uniform sampling density throughout the chromatogram, or as near uniform as possible.

The effect of bunching is similar to changing the sampling time to match the bunch width, but unlike a change in sampling frequency, bunching operations are created by software and can be reversed.


### Baseline Bunching
Over long stretches of stable baseline, stored data samples will merely consist of a series of numbers having the same value. To avoid waste of memory, baseline data are stored in one bunch represented by a single datum and the number of times it recurs.


### Estimating the A/D Sampling Frequency and Bunch Size
Figure 3.15 shows that the minimum number of data samples required to measure a peak area accurately ($<0.1\%$) increases with peak asymmetry. For a symmetrical peak it is about 25,[19] though this is conditional on the $S/N$ ratio being not less than about 25:1. See Figure 3.16.

With greater asymmetry more samples are required. It has been shown[17] that 100 samples/peak gives near maximum peak area accuracy; faster sampling gives negligible improvement.

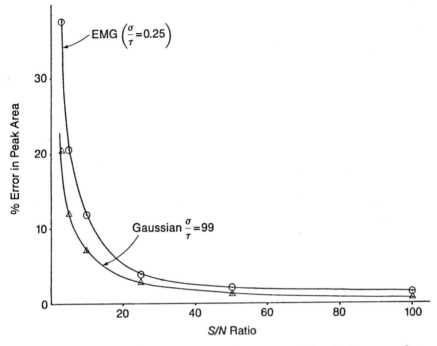

**Figure 3.16** *Accuracy of area measurement increases as S/N ratio improves—faster for symmetrical peaks. Peaks are 10 s base width; sampling interval 1 s (Reproduced with kind permission from J. Chromatogr. Sci., 1988, **26**, 101)*

If a Gaussian peak width is $6\sigma$ (from $-3\sigma$ to $+3\sigma$), then 100 samples/peak is approximately 17 samples/$\sigma$. The standard deviation, $\sigma$, and retention time, $t_R$, are related to column plate number, $N$, by:

$$N = \frac{t_R{}^2}{\sigma^2} \tag{12}$$

or,

$$\sigma = \frac{t_R}{\sqrt{N}} \tag{13}$$

The base width of a Gaussian peak can therefore be expressed in terms of $N$ and $t_R$ by:

$$\text{base width} = 6\sigma = \frac{6t_R}{\sqrt{N}} \tag{14}$$

If this width is to be digitized into 100 samples, the necessary bunched sample frequency, $f$, is given by:

$$f = \frac{100\sqrt{N}}{6t_R} = \frac{17\sqrt{N}}{t_R} \text{ (approximately)} \tag{15}$$

The fixed sampling frequency, $F$, of the A/D converter is designed to sample the fastest peaks, for example early eluters from a WCOT column. If such a peak has a retention time of 250 s (say) in a column of 250000 plates, the sampling frequency necessary to sample this peak 100 times will be:

$$f = \frac{17\sqrt{250000}}{250} = 34 \text{ samples/s} \tag{16}$$

Higher sample densities of 100 samples/peak compensate for asymmetry.

The fixed sampling frequency, $F$, varies with integrator manufacturer in the range of 100 to 1000 samples/s, though this may have to be shared by more than one input channel. It is sufficient for (today's) fast capillary peaks.

Bunching of samples to achieve uniform sample density through the chromatogram is estimated from:

$$\text{Bunching, } B = \frac{F}{f} \tag{17}$$

$$= \frac{F\,t_R}{17\sqrt{N}} \tag{18}$$

During data collection the integrator monitors peak width and issues instructions to vary the bunching to match the peaks being eluted; usually this means increasing the number of samples in a bunch.

In order to keep peak sensing algorithms uniform (the reason for data bunching), the integrator only changes the number of data samples in a bunch between peaks, when the detector signal is on baseline. If there are no suitable stretches of baseline, a queue of bunch update commands can form waiting to be implemented and a uniform sampling density may not be achieved.

**Data Bunching and the Peak Width Parameter**
The number of integrals which constitute a 'bunch' is programmed by the width or peak width parameter. This number is estimated by the analyst when peak width is programmed and represents the initial bunching of data samples. The bunch size is continually reviewed and updated during the analysis by the integrator. If the local peak width exceeds the current bunch size by a prescribed amount the integrator will increase the bunch size accordingly.

Peak width is the most important parameter of the integrator and must be programmed carefully otherwise peaks will be measured incorrectly or even missed altogether. It is not actually expressed in terms of numbers of data samples; for convenience it is related to the widths at half height of early peaks. Peak width has a relatively small range of initial values; a guide to them is given in Table 3.1.

**Table 3.1** *A Guide to Initial Peak (Bunch) Width Values*

| Chromatogram | Peak Width (*Half Height in* ms) | |
|---|---|---|
| WCOTs, Megabore | 100–300 | |
| Microbore, Fast LC | 200–400 | |
| Packed GC | 300–500 | |
| Packed LC | 400–600 | |
| Amino Acid Analyser | 700–1000 | (not LC type) |
| GPC | 1000–20000 | |

Except for GPC, it is unlikely that an initial peak width value greater than 1000 ms will be required. If selection errors are to be made, it is better to err on the small side, *i.e.*, too narrow a bunch, but then the $S/N$ ratio should be good or too much noise will be seen and some area may be lost from the start and end of peaks.

## Peak Width Parameter and Analysis Reprocessing

Reprocessing an analysis must allow the peak width or data bunching to be changed, and the data to be reprocessed with the new value. Adjustments to peak width are normally small from, say, 500 to 400 ms, and since analysis data are stored as integrals, the smallest possible change will be the addition or removal of one datum from the bunched integral.

If changing peak width is a possible requirement, data must be stored at a fast enough sampling frequency to allow the small adjustment. This will require much greater storage capacity, and so the option should be carefully considered.

Once the best choice of peak width has been determined, data can be economically stored at this value on the assumption that no further revision will be necessary.

## Peak Sampling Synchronization

Digitizing a peak signal can be likened to laying a grid over the peak as shown in Figure 3.17.

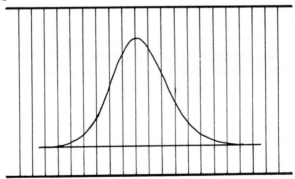

**Figure 3.17** *Peak overlaid with grid*

It should make no difference to the measured peak area where this grid is layed: nudging it to right or left in Figure 3.17 should still give the same area. But it does make a difference; the measured area varies with grid position.[25] Stated another way, measurement of the peak area is not independent of the synchronization of peak start with sampling start.

Lack of uniform synchronization reduces measurement precision, since if it were ever possible deliberately to synchronize the start of one peak with sampling, the combination of regular data sampling and random peak positions would preclude synchronization of other peaks except by chance.

The loss of precision is related to the separation of the peak start from the nearest sampling start, and is therefore reduced by increasing the sampling frequency.

## Rounding or Truncation Errors

Sampling frequency errors have been reported[25] relating to the 'bit resolution' or smallest change in detector signal which the integrator can see, and to the limited number of decimal places used by earlier computer technology.

Errors in 'bit resolution' were reduced with the introduction of auto-ranging A/D converters, and the increased power of modern computers has solved the problem of too few decimal places. Programmers can re-introduce this error, however, by truncating calculations too severely, or at too early a stage during processing.

### Aperture Time Jitter[25]
The precision of the sampling frequency is another problem that more recent electronic technology has reduced to negligible proportions.

## 4   Filtering and Smoothing the Chromatographic Signal

Filtering and smoothing are techniques to suppress or remove noise from the detector signal and to improve the $S/N$ ratio. Too much improves the $S/N$ ratio but distorts peak shape[24,25] leading to a loss of measurement accuracy, height generally suffering more than area. Too little leaves residual noise to blur peaks and interfere with peak measurement, again with loss of accuracy but area suffering more than height. Improvement of the $S/N$ ratio increasingly favours peak area measurement.

### Filtering
Filtering is the real time and irreversible suppression of noise by electronic hardware.

### Smoothing
Smoothing is improvement of the $S/N$ ratio by mathematical operations on

the stored data; it is not irreversible if stored data can be reprocessed. All integrators now process data when the analysis is ended, and data occurring before and after the event can be included in the processing.

The smoothing algorithms used by commercial integrators are unknown to the analyst. He has no control over them or the assumptions they make, except perhaps for some limited selection. All that is known is that the integrator does employ them and such algorithms generally tend to preserve area measurement.

## Electronic Filters

Hardware filters are based on capacitative and inductive noise suppression. 'RC' (capacitative) filters are built into each signal input line. Inductive filters remove spikes from the power supply.

Passive hardware filters are low in cost but fixed in time constant and function. Active filters do the same job but their time constants can be tuned. All hardware filters vary a little with component tolerance.

Common mode rejection (CMR) and cable screening protect integrators against the pick-up of random environmental spikes created by neighbourhood relays switching or heavy machinery powering up. Many laboratories, aware of these problems, already have spike filters built into their power lines to clean up the power supply.

### Sampling Frequency and Mains Coupling

Mains power frequency, 50 or 60 Hz, can couple (add) to the detector signal to some degree. The effect will cancel out over an integral number of cycles as the negative half cycles cancel the positive halves. The fixed electronic sampling frequency, $F$, is therefore selected in part to allow bunching to an integral number of mains cycles.

To avoid manufacturing different instruments for countries with 50 and 60 Hz mains, a sampling frequency of 100 Hz was an early standard because it can be bunched into samples of 10, equivalent to 10 Hz, which is the lowest common multiple of 50 and 60 Hz. Higher sampling frequencies such as 1000 Hz will also bunch into groups which cancel mains pick up. It is a simple filter that costs nothing to manufacture.

## Smoothing

Integrators use software based smoothing techniques which work effectively if the data sampling frequency is constant for the duration of the peak(s) and fast enough to give a large number of samples per peak.

The main smoothing operations are:

(1) digitizing and integrating;
(2) bunching;
(3) polynomial curve fitting (and peak location);
(4) signal subtraction.

These operations work on a single data collection. Unlike many analytical techniques, chromatography rarely generates identical data sets for repeated experiments and ensemble averaging techniques to remove random error have limited value.

### Digitizing and Integrating
Sampling and integrating the detector signal during A/D conversion averages the signal over the sampling interval. It removes noise (and any detailed peak information) occurring inside the interval whose frequency is greater than twice the sampling frequency.

### Bunching
Bunching integrals together further attenuates residual noise by spreading its effect over the width of the bunched group (Figure 3.18)

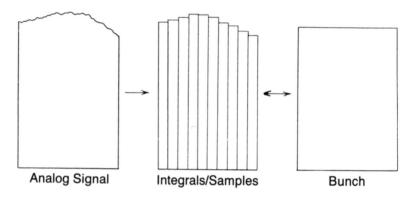

Analog Signal          Integrals/Samples          Bunch

**Figure 3.18**   *Digitizing and bunching time averages noise, but bunching is reversible*

The size of the bunched group is determined by the peak width parameter, and the effect of varying it is to spread noise over greater or lesser time intervals.

Bunching of data samples is a form of boxcar averaging[11,26] in which the number of data gathered into the bunch increases during the analysis to match peak widths and maintain a uniform peak sampling density. No datum is included in more than one bunch. The improvement to $S/N$ ratio increases with the square root of the number of points in the window, $2m+1$:[26]

$$\frac{(S/N)}{(S/N)_o} = \sqrt{(2m+1)} \qquad (19)$$

### Moving Windows and Polynomial Curve Fitting
This third form of signal smoothing also includes simultaneous processing of data to locate peaks.

A moving window scan spanning $2m + 1$ data points, from $-m$ to $+m$ with 0 being the central point, moves along the data replacing the central

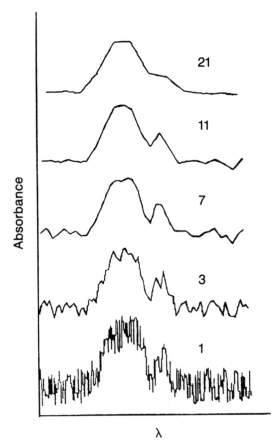

**Figure 3.19**  *Boxcar averaging of data. Number of samples in bunch is shown*
(Reproduced by kind permission of *J. Chem. Educ.*, 1985, **62**, 866)

datum in the window, $y_i$, by the weighted average value of the data within
the window, that is:

$$\bar{y}_i = \sum_{-m}^{+m} C_i y_i / N \qquad (20)$$

where $C_i$ = weighted coefficient, $N$ = normalizing coefficient, and here '='
means 'takes the value of'.

   The window moves forward one point, picking up a new point at the front
of the window, dropping an old one from the back, and averages a new
central datum. The whole data set is processed except for the first and last $m$
data points which are lost and therefore must not include required information.

   Savitsky and Golay[27] published tables of values for $C$ and $N$ which
generated values of $\bar{y}_i$ identical to values which would be produced if the data
had been curve fitted to an $n$th order polynomial by linear regression. The
polynomials range in order from 2nd to 5th and the number of data points in
the smoothing window range from 5 to 25. Additional tables allow calcu-

lation of the 1st to 5th derivatives of the curves. The 1st and 2nd are used for
peak sensing and location; higher derivatives can be similarly used if the $S/N$
ratio is high enough.

Savitsky–Golay smoothing assumes the data sampling rate to be constant
and the noise to be random,[28] ideally with a much higher frequency to the
peaks. Improvement in the $S/N$ ratio increases with the number of data in
the moving window $[\sqrt{(2m+1)}]$ but so does peak distortion. The tables of
coefficients are symmetrical and really apply only to symmetrical peaks. If
peaks are asymmetric to begin with, smoothing generates further distortion,
peak broadening and loss of peak height. The worse the initial asymmetry,
the greater the distortion, but it is reduced if smaller windows and higher
polynomial orders are selected.[29]

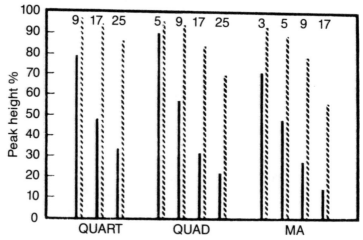

**Figure 3.20**   *Moving window smoothing (Savitsky–Golay) produces shape distortion
if sampling frequency is not high enough. Symmetrical peaks - - -;
asymmetric peaks ——*
(Reproduced with kind permission from *J. Chromatogr.*, 1976, **126**,
279)

Smoothing distortion is reduced to a negligible amount if the sampling
density is high enough and the ratio of window/peak width is small, but the
time required to perform the calculations increases (Figure 3.21). Peak
distortion becomes apparent when the window size exceeds about 67% of
the peak half width.[28]

Savitsky–Golay tables, later corrected by Steiner[30] for some numerical
errors,* still form the basis of integrator smoothing and peak detection.
Manufacturers have improved them, though little is published as the details
have commercial value.

## Optimum Filtering

If a narrow filter leaves too much noise, and a broad one produces peak
distortion, it poses the question of what is the optimum filter width.

---

*Not all of the errors were reported by Steiner. For example, on p. 1637 of Savitsky and
Golay's paper, on Repeated Convolution, $d_{-3}$ is $-21$ not $-33$ and $d_3$ is $+21$.

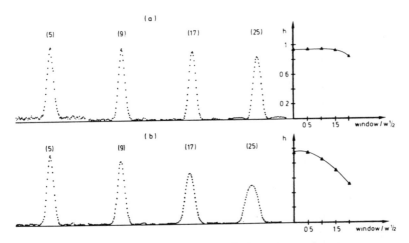

**Figure 3.21** *Distortion, $h/h_o$, of a Gaussian peak as a function of the ratio between window size and half-height width for* (a) *polynomial smoothing and* (b) *moving average. The window size $(2m+1)$ is indicated in parentheses* (Reproduced with kind permission from 'Chemometrics, a Textbook', Elsevier, Amsterdam, 1988)

Van Rijswick[31] developed an equation based on a Gaussian peak, relating the $S/N$ ratio before and after filtering. He showed that:

$$\left(\frac{S}{N}\right)_{\text{filt}} = 2.174 \left(\frac{S}{N}\right)_{\text{orig}} \cdot \sqrt{\left(\frac{w_p}{\Delta t}\right) \cdot \left[\frac{K^5}{(1 + K^2)^3}\right]^{\frac{1}{2}}} \tag{21}$$

where $w_p$ = half width at inflection height before filtering (for Gaussian peak, $w_p = \sigma$)

$N$ = average noise amplitude before filtering

$\Delta t$ = sampling interval

$$K = \frac{\text{filter width}}{\text{peak width}} = \frac{2m+1}{w_p}$$

Optimum filtering is found by optimizing the $K$ term, which has a maximum value at $K = \sqrt{5}$ (Figure 3.22).

Interpreting this for a Gaussian peak, the optimum filter width is $\sqrt{5}\,\sigma$, and if this peak is sampled 100 times (or 17 samples/$\sigma$), the width of the filter is 38 data points, which would require considerable computation. Fortunately, Figure 3.21 has a broad enough maximum and a smaller value of $K$ can be used without much penalty. An additional consequence of filtering is an increase in peak width to $\sqrt{[w_p^2 + (2m+1)^2]}$.

Cram[29] showed that where peak asymmetry is present, it is necessary to use narrow windows with 4th and 5th order polynomial smoothing rather than broad windows with 2nd and 3rd order polynomials or moving average to minimize the additional distortion from the smoothing operation.

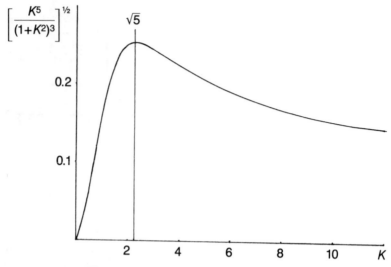

**Figure 3.22**   *Location of optimum filtering*

All chromatographic peak smoothing ends up as a compromise between residual noise and peak distortion.

## Signal Subtraction

A recent form of signal processing made possible by the continuing developments in microcomputer technology involves the removal by subtraction of recurring events or systematic error, such as drifting baselines, valve switching spikes, *etc*.

Essentially two chromatograms are stored, the analysis and the baseline, *i.e.* the analysis without solute injection. Provided that the analysis is repeatable, subtraction of the second from the first removes baseline disturbances common to both, and gives a chromatogram with the flat baseline.

There is a price: the difference chromatogram combines the inaccuracies and imprecision of both originals.

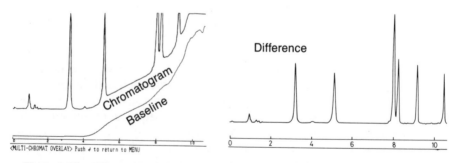

**Figure 3.23**   *Chromatogram and baseline are stored separately. The difference chromatogram removes baseline drift*
(Reproduced by kind permission of Shimadzu)

Subtracting a wandering baseline in this manner is one justifiable way to remove a curved baseline from beneath a group of peaks; at other times integrators construct straight baselines beneath peaks. This principle is extended to allow comparison of chromatograms: one chromatogram can be subtracted from another to highlight differences between the two.

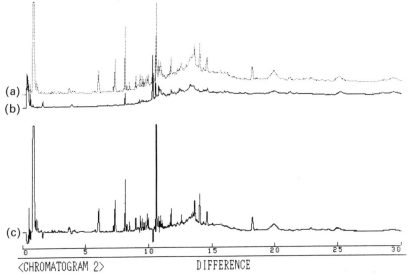

**Figure 3.24** *Subtraction of one chromatogram* (a) *from another* (b) *to give difference* (c)
(Reproduced by kind permission of Shimadzu)

## Chromatogram Plotting

The stored signal data are passed through a D/A converter in order to reconstruct and plot the 'original' chromatogram. If the plotter is more than a mere dot matrix printer, the reconstruction will give no hint of the digital processing which has taken place. If the integrator or computer offers screen expansion, however, it is possible to magnify the peaks and see the individual data samples represented as simple ordinates joined together for screen reconstruction (Figures 3.25 and 3.26).

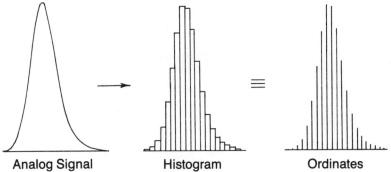

**Analog Signal**          **Histogram**          **Ordinates**

**Figure 3.25** *Digitization of an analog signal to give a series of ordinate values*

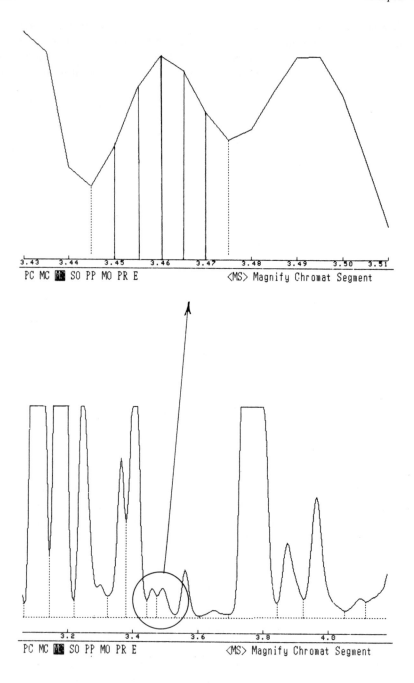

**Figure 3.26**  *Expansion of digitized signal shows ordinates joined to display a peak*

# 5  Location and Measurement of Peaks

Accurate peak location is made by coarse and fine scans:

(1) approximate peak location to detect every peak is obtained from Savitsky–Golay first and second derivatives;

(2) precise location of each peak start and end point, the limits of integration, is obtained by a local search.

## Finding the Peaks

Any peak can be divided into four quadrants by lines drawn through the peak maximum and points of inflection (Figure 3.27).

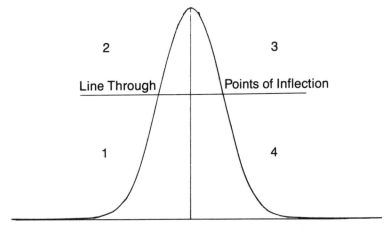

**Figure 3.27**  *Symmetrical Gaussian peak*

Each quadrant is uniquely characterized by its first and second derivatives and this is used as a coarse peak locator and shape test (Table 3.2).

**Table 3.2**  *First and Second Derivatives of Each Quadrant of a Peak*

| Quadrant | $dy/dt$ | $d^2y/dt^2$ |
|----------|---------|-------------|
| 1 | $>0$ | $>0$ |
| 2 | $>0$ | $<0$ |
| 3 | $<0$ | $>0$ |
| 4 | $<0$ | $<0$ |

The first and second derivatives are obtained from Savitsky–Golay convolution of the stored data. Peaks are located where the first derivative departs from zero by more than a prescribed threshold, the slope sensitivity $S$ (Figure 3.28).

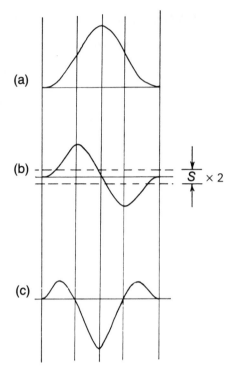

**Figure 3.28**  *Peak, first and second derivatives*

The non-zero threshold distinguishes between the relatively small baseline gradients and the larger peak gradients. Slope sensitivity reduces false peak detection, but inhibits precise location of the limits of integration. Nevertheless, approximate peak location shows where the peak boundaries are to be found and later prevents the integrator making detailed searches on stretches of empty baseline, and so reduces the processing time. If the $S/N$ ratio is large and the baseline is flat, slope sensitivity can be given a small value, and the peak limits will be located with good precision even in the 'coarse' search.

**Retention Time**
The peak maximum is located where the first derivative is zero between peak boundaries (Figure 3.28) and from this the retention time is measured. Without curve fitting, retention time accuracy and precision would be determined by the bunched sample interval.

**Peak Shape Test**
The combination of values given in Table 3.2 allows the integrator to examine the peak for shape and enables the integrator to distinguish between peaks and baseline ramps caused by solvent gradients or temperature programs. It is a measure against false peak detection.

## Properties of the Smoothed Data

Savitsky–Golay processing to locate the first derivative is equivalent to the integrator comparing each smoothed datum to its $M$th predecessor where $M$ is the width of the convolution window ($= 2m+1$).

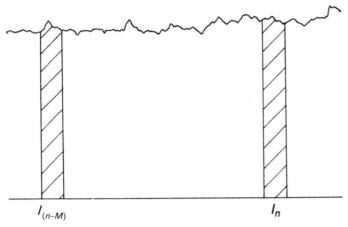

**Figure 3.29** *The integrator constantly computes* $\Delta I_n = I_n - I_{(n-M)}$

## Data Integrals

If $I_n$ is the most recent smoothed datum or integral, the integrator computes:

$$\Delta I_n = I_n - I_{(n-M)} \tag{24}$$

and the value of $\Delta I_n$ is positive, negative or 'zero' (as defined by slope sensitivity) depending on whether the baseline is rising, falling or is flat. $\Delta I_n$ is therefore monitoring slope and its value, which is a measure of the first derivative, can be used to show when a peak is emerging ($\Delta I_n > S$), or when the signal is on baseline ($-S < \Delta I_n < S$). Integrators which have slope indicators, keyboard LEDs for example, have them controlled by this quantity and they switch on and off according to its value.

The integrals are used in three ways as described below.

### (1) As an Area

Integrals are measured in units of $\mu V$ sec. When those within peak limits are summed and corrected for baseline, they will compute the peak area.

### (2) As a Height

Since the integral's value is determined for a fixed time interval against a constant reference level, $-5$ or $-10$ mV, it can only vary if the baseline rises or falls. It is therefore entirely height dependent.

To maintain this, integrals must not change their bunch size during peak measurement, and 'time to double' instructions are only carried out after the peak end-point has been detected, when the signal is back on baseline.

### (3) As a Gradient

$\Delta I_n$ is not only a difference in baseline height, it is a difference in height measured over the fixed time interval of $M$ data slices, and so $\Delta I_n$ is also a gradient (Figure 3.30). This gradient has a discontinuity when the bunching is changed on baseline, but the integrator is programmed to account for it.

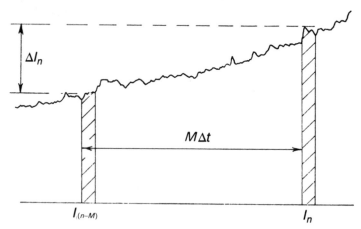

**Figure 3.30**  *$\Delta I_n$ is measured over a fixed time interval of $M$ samples*

### Baseline Fluctuations and Slope Sensitivity

Slope sensitivity is determined by giving $\Delta I_n$ a threshold value. It is the smallest shift in baseline position which the integrator sees as positive or negative slope.

### Programming Slope Sensitivity

Judging what value to give to slope sensitivity is not intuitive. Integrators therefore provide a test facility to monitor baseline for a while, measure the average value of signal fluctuations, and self-program a value. The length of the test is linked to the peak width setting and lasts for approximately 100 bunched integrals.

Daily use of the test should make the analyst familiar with the empirical range of values to be expected; different columns, detectors, and sample types all contribute to the range. With experience, the analyst may even choose to over-rule the integrator and enter an alternative value for slope sensitivity, for example, to allow more micro peaks to be seen if these are important.

### Slope Sensitivity and Representative Baseline

Automatic tests to determine baseline noise levels are performed when the signal is on 'representative' baseline.

If an integrator is asked to measure baseline noise following a thorough reconditioning of the column, it will measure a different value to that produced after samples have been analysed and the baseline is littered with chromatographic rubble. If this latter baseline is the normal one, it will be more appropriate to determine a slope sensitivity value which includes the baseline rubble and considers it to be 'zero slope'.

If programming errors are to be made, it is better to make slope sensitivity too sensitive. The integrator will see more noise and measure too many small peaks, but these can be removed from the final report by setting a minimum area threshold.

### Updating Peak Width and Slope Sensitivity during Analyses

As analyses progress peaks become broader and flatter. The initial values for peak width and slope sensitivity require revision and updating. Integrators do this automatically by monitoring peak widths and comparing them to the programmed value. If there is a significant difference, the peak width parameter is increased (*i.e.* the sample bunching is increased).

Alternatively the parameters may be programmed to increase on a timed basis via a time program.

When bunching is changed it changes the values of both the peak width parameter and the slope sensitivity (Figure 3.31).

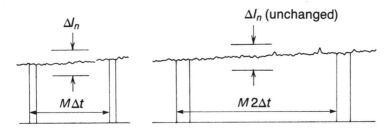

**Figure 3.31** *If peak width parameter is doubled, $\Delta I_n$ is measured over twice the time interval and slope sensitivity is halved*

If the bunching is doubled the effective width of each data integral is doubled, and $\Delta I_n$ is measured over twice the time interval while still recognizing the same shift in baseline level as the slope threshold. Expressed as a gradient, slope sensitivity is halved and becomes twice as sensitive.

## Location of the Limits of Integration

Figure 3.32 shows a single peak on a flat baseline, sliced into bunched integrals as determined by the peak width parameter, and a reference level above which the detector signal is measured. The reference level is the lower limit of the integrator's operating range ($-5\,\text{mV}$ or $-10\,\text{mV}$). If the detector signal drifts lower, or is inadvertently sent lower by careless use of detector back off control, it will create a dead band error and only that part of the detector signal which projects above the reference level will be measured.

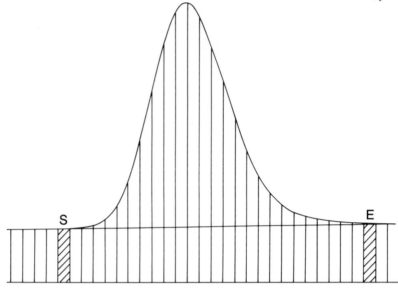

**Figure 3.32**  *Peak start*, S, *and end*, E

The approximate peak start is located where the detector signal first exceeded the programmed slope sensitivity value. The integrator proceeds to conduct a fine search for a more accurate location among the preceding data samples. The range of its search is limited by the time interval between first peak location, S, and the peak maximum. This is no more than $2\sigma$, where $\sigma$ is the peak standard deviation; the integrator looks back over the range $4\sigma$ from the peak retention time. The 'correct' peak start is located at the lowest baseline level in this region.

As the peak emerges, the integrals increase in value compared to their predecessors (Figure 3.32).

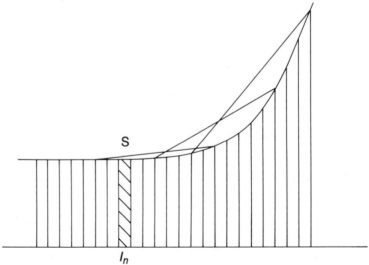

$I_n$

**Figure 3.33**  *The peak starts at* S *when* $\Delta I_n$ > *slope sensitivity*

To be recognized as the start of a peak and to trigger integration, $\Delta I_n$ must be positive and increasing in value for $P$ consecutive integrals, where $P$ has a value which depends on the moving window size (and manufacturer) but ranges from about 5 to 10. If it is too large the integrator will be slow to make decisions; if it is too small the integrator will measure residual noise peaks.

When the integrator observes $P$ positive values of $\Delta I_n$ in succession, such that:

$$\Delta I_n > S \tag{25}$$

$$\text{and } \Delta I_{n+1} > \Delta I_n \tag{26}$$

$$\text{and } \Delta I_{n+2} > \Delta I_{n+1} \tag{27}$$

$$\vdots$$

$$\text{and } \Delta I_{n+P} > \Delta I_{n+P-1} \tag{28}$$

it satisfies the conditions of the peak shape test that $dy/dt$ and $d^2y/dt^2$ are both positive. The integrator triggers integrate mode which initiates several operations:

(1) Data samples are accumulated from the peak start.
(2) A number of baseline integrals immediately before the peak start are averaged and stored.
(3) A peak event mark (if available) will be added to the chromatogram. When integrators and strip chart recorders were separate instruments, the event mark added by the integrator was drawn late by the recorder, at the point where the peak start was confirmed, not where peak onset was first noted. Just how late was determined by the peak width value. Now that intelligent plotters are part of the integrator, printing of the chromatogram is delayed slightly so that the event mark can be added to the correct place, but this has sacrificed a useful indicator of whether the peak width parameter has been programmed correctly.
(4) Integrators with keyboard indicators such as LED or LCD display flag when a peak is confirmed.

**Small Peak Filtering**
Unless small peaks show positive slope for the correct length of time, $P \times \Delta t$, they will not trigger integration and will not be measured. Varying the bunch width $(\Delta t)$ varies the qualifying time, allowing selective filtration of small peaks. This defines the fastest measurable peak.

## Location of Peak End

The integrator continues to accumulate peak integrals until it locates the baseline again.

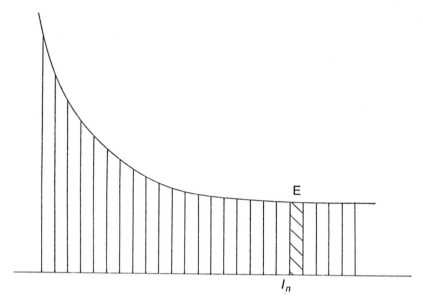

**Figure 3.34** *Location of peak end*

The approximate location of peak end is determined when the detector signal gradient, $\Delta I_n$, becomes zero for a qualifying time of $Z$ consecutive samples. $Z$ is larger than $P$ because of peak tailing which has to be monitored longer.* The integrator conducts a fine search among the successive data samples to find a more accurate location. The search range is determined by the time interval from the peak maximum to the approximate end, which will be large if there is tailing. The correct end point of the peak is taken to be the lowest baseline point in the search region.

When baseline is confirmed, more operations come into play:

(1) Integrate mode is terminated and accumulation of peak integrals stops at the peak end. The integrator now holds in its memory the sum of the data points from peak start to end (Figure 3.35).

(2) Several of the new baseline integrals immediately after the peak are averaged and stored.

(3) A peak event mark is recorded on the chromatogram, and keyboard peak indicators switch off.

## Measurement of Peak Area

Peak area measurement is completed by subtracting the 'baseline area' from the accumulated integral count (Figure 3.36). The baseline area is calculated from the average of the stored baseline integrals before and after the peak, and the peak width (from the number of integrals accumulated within the peak limits).

*On at least one integrator, $Z = 2P$.

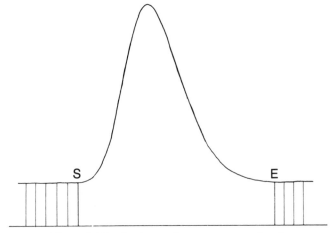

**Figure 3.35** *Peak area between start and end*

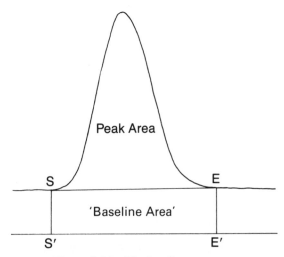

**Figure 3.36** *The baseline area*

The baseline area is calculated as a trapezium, and the method is commonly called 'trapezoidal baseline correction'.
There are three points to note:

(1) Sloping baseline. It makes no difference whether the baseline is sloping or not. By calculating the baseline area beneath the peak as a trapezium, the integrator constructs a linear baseline which joins the beginning and the end of the peak and there is no assumption that these points are at the same level (Figure 3.37).

(2) The interpolated baseline is a straight line. The integrator draws the same baseline that the chromatographer would draw with a pencil and ruler.

(3) The only assumptions made about peak shape are that the signal has gone up and come down again within time limits allowed by $P$ and $Z$.

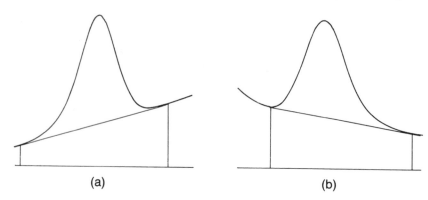

**Figure 3.37** *Trapezoidal baseline correction compensates for a shift in baseline level*

### Baseline Convention

Integrators construct linear baselines under peaks though in many cases the real baseline is known to be curved.

It is a convention from the days of manual measurement that a reproducible straight line is better than a subjective curve simulation; no two chromatographers would draw exactly the same curve. After more than two decades of search, there is still no easy formula to construct the true baseline beneath a peak. There is the single exception of storing and subtracting nonlinear baselines from a blank run.

The use of logarithmic or exponential models of baseline can be given general theoretical justification but it is difficult to justify fitting a curve to a specific peak if it can vary in shape from one analysis to the next.

## Measurement of Peak Height

The procedures for measuring peak height are very similar to those for peak area. The difference is that, instead of accumulating peak integrals, the integrator selects, as an approximation, the largest value (Figure 3.38) and fits the data in that region to a curve from which the height may calculated. Trapezoidal baseline correction of this height gives the reported peak height.

## Measurement of Retention Time

Retention time is measured as the time of the computed peak height. This is equivalent to determining when $\Delta I_n$ changes in value from positive to negative while in integrate mode.

The integrator monitors the trend in slope before confirming the peak maximum so the signal has moved past the maximum before the change is noted. On real time chromatograms the integrator prints the retention time after the peak maximum has eluted, when the detector signal is falling.

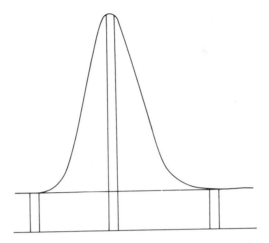

**Figure 3.38**  *For peak height measurement the integrator selects the largest smoothed datum*

## Measurement of Two Unresolved Peaks

As far as the integrator is concerned, the only difference between a fused pair of peaks and a single peak is the valley between them (Figure 3.39).

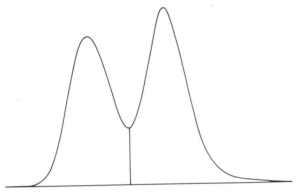

**Figure 3.39**  *Fused peaks*

The integrator locates the correct start of the first peak by means of a fine scan of the data and begins there to accumulate. data until it locates the valley.

A **valley** is defined where the detector signal changes slope from negative to positive for $P$ consecutive integrals while in integrate mode. It is approximately located where the Savitsky–Golay first derivative is zero between negative and positive slope values, and precisely located in this region by finding the curve minimum.

At the valley, accumulation of data samples from the first peak is ended and the total stored. Accumulation of the data samples of the second peak begins. The valley position determined from the curve fit does not in general coincide with the edge of a data sample, so the data sample which straddles the curve minimum is partitioned to each peak (Figure 3.40).

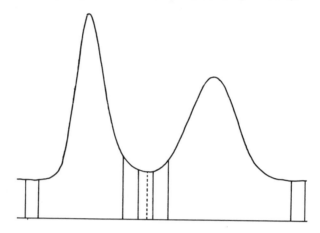

**Figure 3.40**   *Partitioning of data sample at curve minimum*

The peak end is located, as for a single peak, in a fine search beyond the approximate location, and accumulation of data for the second peak is terminated.

During this time peak event marks and retention times have been added to the chromatogram and the keyboard flags will have indicated the peaks' presence.

**Measurement of Peak Area**

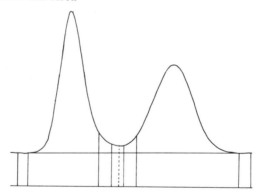

**Figure 3.41**   *'Baseline area' for fused peaks*

The trapezoidal 'baseline area' beneath both peaks is computed from the average of the stored baseline integrals before and after the peaks and the base width of the peaks. The area is divided at the valley into the two parts corresponding to the areas below each peak, and each peak area is calculated by subtracting each baseline area from the peak total (Figure 3.41).

### Peak Measurement Diagnostics

In the Final Report the integrator highlights peaks which have been separated by perpendiculars by printing a diagnostic code alongside the peak area. The code varies with manufacturer but is usually a letter like P (perpendicular) or V (valley); sometimes it is a numeric code.

## Measurement of Fused Groups

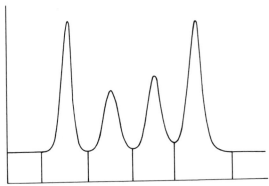

**Figure 3.42**   *A group of fused peaks*

Larger groups of unresolved peaks simply have more valleys. The start of the first peak and the end of the last are measured as in the case of a single peak. Peak integrals up to each valley are summed and stored and the peak retention times are measured.

The trapezoidal baseline area beneath the whole group is measured, and divided at the valleys into the baseline areas below each peak; these individual baseline areas are subtracted from the peak integrals to yield each peak area.

None of the peak area measurements can be completed until baseline is located at the end of the group and trapezoidal baseline correction is made. In WCOT 'forests' the whole of the chromatogram may be in temporary storage until the signal returns to baseline at the end of the analysis.

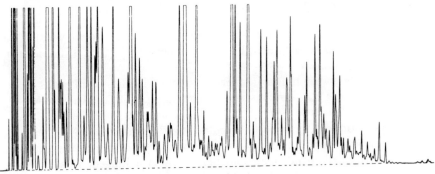

**Figure 3.43**   *In a WCOT forest the area of the first peak cannot be measured before the baseline is found at the end of the chromatogram*

## Shoulders

A shoulder is an unresolved peak on the leading or trailing edge of a larger peak. There is no true valley in the sense of negative slope followed by positive slope.

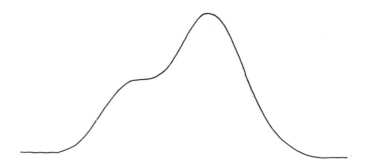

**Figure 3.44**   *Varying slope sensitivity can force shoulder recognition*

Integrators define shoulders as regions of zero slope between two positive or two negative slopes, while in integrate mode.

What actually counts as 'zero' slope is determined by slope sensitivity, and relatively steep shoulder slopes can be made to be 'zero' by temporarily increasing the value of slope sensitivity to force shoulder recognition and peak detection.

When it finds a shoulder, the integrator will process it as a valley and drop a perpendicular from the point of minimum slope.

Shoulder measurements are not accurate. The area of the shoulder, or smaller peak, is over-estimated[32] and includes part of the larger peak. Both areas are inaccurate. No important peak should be a shoulder; separation must be improved. If this is not possible, conclusions drawn from shoulder measurements should be used with caution.

## Measurement of Tangent Peaks

At each valley, integrators perform an additional size test to see whether the next peak should be separated by a perpendicular or whether it should be skimmed by a tangent.

Historically and commonly, tangent skimming implies small peaks on the tails of big peaks but some integrators offer the facility to skim the leading edge too.

The decision to skim or not is based on relative peak size and the height of the valley between the peaks. A size threshold is set above which peaks are separated by a perpendicular, and below which a tangent is used. Some integrators can vary this threshold, others have it fixed with provision to override it and force separation exclusively by perpendicular or by tangent.

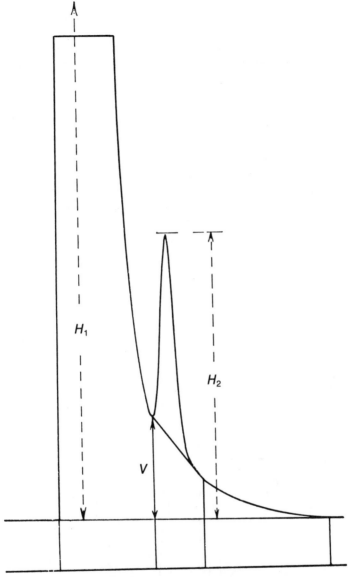

**Figure 3.45** *Tangent skimming*

A tangent is skimmed if:

$$\frac{H_1 - V}{H_2 - V} > \text{Size ratio} \tag{29}$$

If not, the peaks will be separated by perpendicular cleavage.

The tangent is drawn from the valley before the small peak to that point after it where the detector signal gradient $(\Delta I_n)$ is equal to the tangent gradient. The size ratio is usually about 10 and not as numerically critical as might first be thought. It only determines whether a particular peak should be skimmed: it has no influence on the peak measurement once this decision is made. Provided the integrator skims those peaks which the chromatographer expects to see skimmed, the value of the size ratio will not be questioned. (See also p. 55.)

**Tangent Diagnostics**
Any peak which has been tangent skimmed off another peak will have a diagnostic code, *e.g.* 'T', printed against it in the final report.

**Tangent Skimmed Groups**
If a group of fused peaks is situated on the tail of a larger peak, the size ratio test may decree that the whole group is skimmed from the larger peak but separated from each other by perpendiculars dropped to the tangent drawn beneath the group.

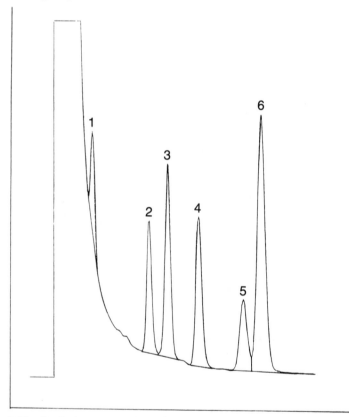

**Figure 3.46**   *Peaks 1 to 6 are skimmed on size comparison with the solvent. Peak pairs 2 and 3, and 5 and 6 are separated by perpendiculars on size comparison with each other*

In the final printed report, each peak in the group will have a diagnostic code indicating that it was skimmed from the larger peak and separated from the other skimmed peaks by perpendicular.

## Small Peaks between Larger Ones

If there are small peaks between larger ones it is perfectly possible that some of them will be selected for skimming while others are separated by perpendiculars. Where the tangent peak straddles the valley position between two larger peaks, it prevents location of the valley which is then shifted to the nearer edge of the smaller peak.

There is no integrator solution to this problem; the only solution is improved peak resolution.

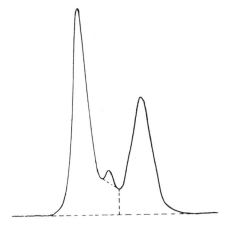

**Figure 3.47**   *A small peak between larger ones*

## Tangents on Tangents

It is possible, though in practice unlikely, that three consecutive peaks might have the appropriate relative sizes to cause the second peak to be skimmed from the first, and the third peak to be skimmed from the second.

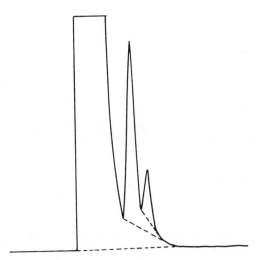

**Figure 3.48**   *Tangents on tangents*

# 6   Baselines, a More Detailed Discussion

An integrator monitoring an analysis and not subject to an integrate inhibit instruction, is either 'on baseline' or in 'integrate' mode. When it is on baseline it will continue to recognize baseline until it identifies the start of a peak and enters integrate mode. It cannot leave integrate mode until it relocates, or is instructed to locate baseline.

Baselines have been hitherto described as 'zero slope' within the tolerance allowed by slope sensitivity, but this is an inadequate definition. Real baselines can slope as a result of solvent or temperature programming, and there are many parts of a chromatogram which show zero slope but are not baseline, for example, valleys and the tops of peaks. The tail of a solvent peak is baseline for the small peaks riding it, yet it may be falling almost vertically. Figure 3.49 shows some examples.

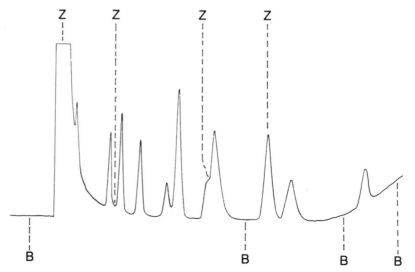

**Figure 3.49**   *Not all baseline is zero slope, not all zero slope is baseline. Z = zero slope; B = baseline*

Positive sloping baselines are a greater problem than negative ones since unwanted or mis-timed peak detection will occur. Some positive slope can be tolerated within the definition of slope sensitivity but this does not prevent the integrator from finding 'zero' slope at peak tops or in valleys if they last long enough to qualify as baseline, *i.e.* for Z consecutive samples.

## Baseline Drift Limit

A better definition of baseline limits the regions where it may be established to where the analyst expects to find it. A second parameter is required to define the boundaries of baseline position and say how far the baseline position may shift during a peak or group. The parameter has a variety of names from different manufacturers but is a 'baseline drift limit' which comes into play only when a peak is detected. Above the limit, zero slope will not be recognized as baseline no matter how long it lasts.

There are two possible ways to define limit: as a fixed millivolt threshold, or as a gradient which increases the allowed range with time (Figure 3.50).

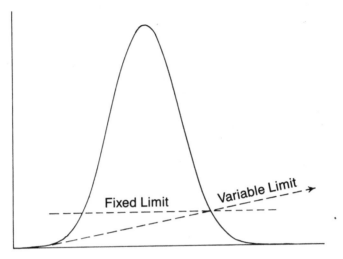

**Figure 3.50** *Limiting the range of baseline drift to below the dashed line*

The fixed threshold cannot account for unexpected shifts in baseline which will lock the integrator into integrate mode until the end of the analysis and measure peak areas down to the pre-shift baseline.

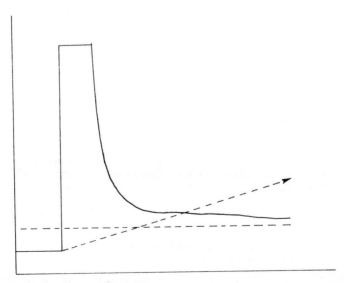

**Figure 3.51** *A fixed baseline limit cannot cope with excessive tailing*

A rising limit eventually allows the correct baseline to be established. Peaks measured before the return to baseline may be measured incorrectly, but later peaks will be preserved.

Neither concept is perfect. Stored analyses can be reprocessed with corrections provided that the analyst is aware of what has happened. Integrators which display the baseline on chart or VDU are a great help in this respect.

**False Starts**

One function of a baseline drift limit is to prevent integration of positive sloping baselines as peaks. Integration is automatically terminated by the integrator when the signal levels off within the allowed baseline range for the qualifying time.

False starts are common on noisy baselines; associated event marks litter the baseline and confuse interpretation (Figure 3.52).

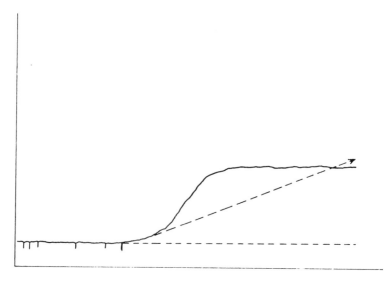

**Figure 3.52**  *Shifts in baseline can be 'seen' as peaks*

## Formulating a Baseline Definition

There is no simple neat definition of baseline. Integrators use a selection of definitions which can be invoked collectively or individually. To an integrator, baseline is defined as:

(1) the detector signal encountered immediately after pressing 'start' (of analysis);
(2) the detector signal encountered immediately after the end of an 'integrate inhibit' instruction;
(3) after a peak, zero slope within slope sensitivity, lasting for $Z$ consecutive data samples below the baseline drift limit;

(4) wherever the analyst forces the integrator to ignore its own logic and establish baseline;

(5) the detector signal at the end of analysis, *i.e.* the last stored datum.

The consequences and short-comings of these definitions are best illustrated by example as shown below.

## Mis-timing 'Start' and 'End'

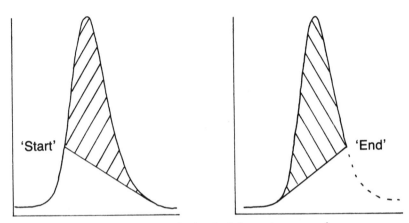

**Figure 3.53** *Only the shaded area is measured*

The shaded areas of Figure 3.53 show the effect of starting the analysis too late or ending too early. The measured areas are incomplete. This type of error happens if automated analyses lose synchronization between injection time and integrator start (Figure 3.54).

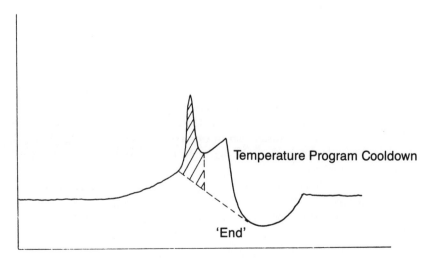

**Figure 3.54** *Analysis ends late, after cooldown has started*

## End of 'Integrate Inhibit'

Integrate inhibit is a programmable function the analyst uses to prevent the measurement of a real but unwanted peak such as the solvent, or to prevent baseline disturbances from interfering with peak measurement.

The end of integrate inhibit is timed to establish a section of true baseline before a peak of interest emerges (Figure 3.55).

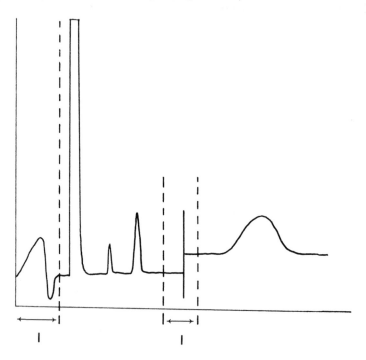

**Figure 3.55** *Integrate Inhibit suspends peak recognition*

## Forcing Baseline

The purpose of forcing baseline is to rescue a peak from incorrect measurement resulting from baseline disturbances. If the baseline near a peak is disturbed by a negative dip or a pressure pulse it will cause at least one integration limit to be established in the wrong place, cause the trapezoidal baseline beneath the peak to be wrongly placed and the peak area to be measured incorrectly.

Baseline disturbances too close to a peak will distort the peak shape and preclude accurate measurement. Forcing baseline is only sensible when there is a small stretch of real baseline on either side of the peak to latch on to. If there is none, further method development to distance the peak from the disturbance is essential.

Forced baseline can also direct the integrator to draw a baseline between points on either side of the peak, which, in the analyst's judgement, will form an accurate trapezoidal baseline under the peak and allow correct measurement.

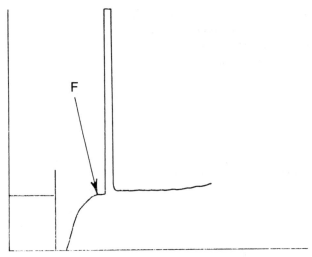

**Figure 3.56** *Force baseline at F*

Integrate Inhibit and Forced Baseline often have the same role.

## Incorrect Programming of Baseline Drift Tolerance

Low valleys between nearly resolved peaks are always a problem. The drift tolerance must be programmed to recognize the lowest valley in the chromatogram (Figure 3.57).

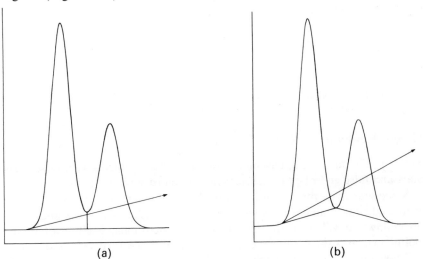

(a)                 (b)

**Figure 3.57** (a) *The valley is outside of baseline range and remains a valley.* (b) *The integrator locates zero slope within baseline range inside the valley and establishes baseline*

The analyst can check that the integrator has correctly found a valley by referring to the VDU display or diagnostic codes printed alongside the peak areas in the final report. When a valley is found, a valley diagnostic will be printed. If no valley code is there, it is because the integrator has considered the valley to be a baseline point and so has no valley to report.

Unless the integrator draws the baseline on the chromatogram or shows it on a VDU, the analyst has only the diagnostic code to confirm correct measurement. The ratio of the peak areas will not differ much between Figures 3.57(a) and (b) and cannot be used as a diagnostic.

## Fused Tangent Measurement

A solvent tail is baseline to the peaks which ride on it. The valley between fused riders may lie below the start of the first rider peak, and require a *negative* baseline limit to ensure that the valley is not recognized as a baseline point (Figure 3.58a). This is normally done automatically by the integrator but if the baseline drift tolerance has been programmed by the analyst, it may not be automatic.

The analyst must check the VDU or the diagnostics allocated to these peaks to confirm that a valley has been correctly identified between them.

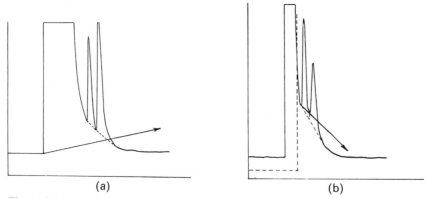

|    (a)    |    (b)    |

**Figure 3.58**   (a) *Fused tangent measurement.* (b) *If the solvent peak is ignored by integrate inhibit, a negative limit is required for rider peaks*

## Single Peaks on a Rising Baseline

Single peaks on a rising baseline have additional problems with false early start and false end of peak measurement (Figure 3.59).

Slope sensitivity must be tolerant of the baseline gradient but pick up the increased peak gradient as positive slope. The baseline drift tolerance must also allow the 'valley' at the end of the peak to establish itself as a baseline point—which is the complete opposite of what is required for fused peaks.

Peaks on fast rising baselines are difficult for any integrator to deal with, especially if the analysis is not easy to repeat, and baseline points forced on a time basis cannot be relied on.

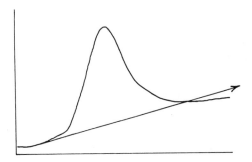

**Figure 3.59** *A single peak on a rising baseline*

The analyst should attempt to improve the chromatography by reducing the baseline gradient or improving retention time stability. If this cannot be done, there is no alternative but to store and reprocess each analysis in turn for the best results.

## Valleys between Fully Resolved Peaks

It is possible for integrators to drop a perpendicular of zero length (Figure 3.60).

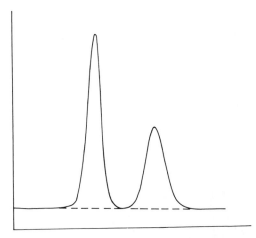

**Figure 3.60** *Fully resolved peaks with a valley between*

The peaks in Figure 3.60 are only just fully resolved but there is no qualifying baseline between them. If baseline drift is restricted so that a valley is forced, the integrator will drop a perpendicular of zero length. With or without baseline detection the area is the same.

## Negative Dips and Constructed Baselines

Negative dips in the detector signal, arising from negative peaks, injection pulses, *etc.*, are major sources of error in baseline assignment. When the signal dives into a trough the integrator will track it, at least as far as the lower limit of its input range. When the signal returns towards its original level, peak detection is triggered on the positive slope and original baseline position is measured from the bottom of the trough. True baseline is only relocated when the baseline drift parameter returns slowly from the trough. The constructed baseline and the drift vector are one.

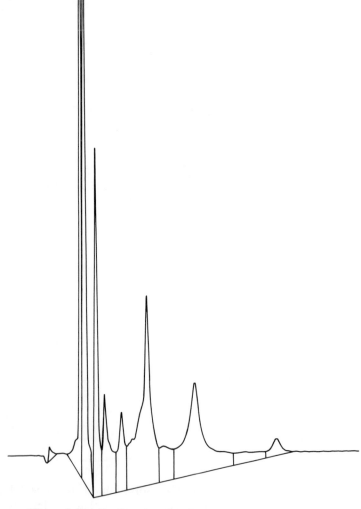

**Figure 3.61**   *Negative baseline dips distort baseline allocation*

The depth of the trough and the relatively slow rise of the baseline boundary create an artificial baseline above which peaks are measured. Very small peaks are given incongruously large areas because of the area below the true baseline (Figure 3.61).

If a series of small peaks, all roughly the same size, have areas decreasing in size away from a baseline dip it is a sure sign that the baseline has been taken from the bottom of the trough (Figure 3.62).

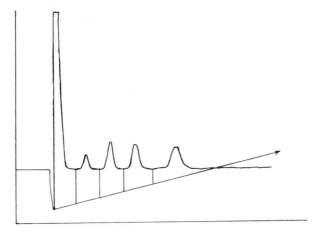

**Figure 3.62** *Baseline taken from the bottom of a trough. Baseline drift vector **is** the baseline*

## Assigning Baseline beneath the Whole Chromatogram

On many isothermal or isocratic chromatograms a line drawn from the start to the end of the analysis will underscore the trace so that no part of the chromatogram descends below. This line can be regarded as the true baseline (Figure 3.63).

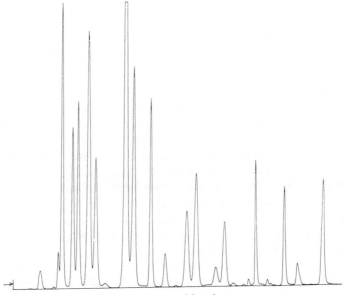

**Figure 3.63** *Level baseline*

Where the baseline is sloping, the integrator will assign straight baselines under single peaks and under each fused group. If the shift in baseline across a group is large, some valleys within the group may project below this construction.

An integrator draws the baseline below a group of peaks in stages. The first construction will join the start of the first peak to the end of the last peak with a straight line (Figure 3.64) provided the end point is within the range allowed by the baseline drift parameter. If no valley points intersect this baseline no amendments are necessary and it will be used in the measurement of peak areas.

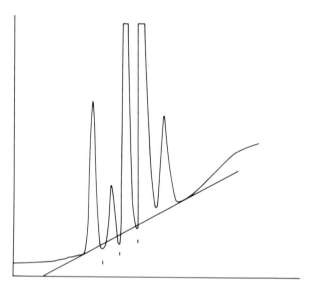

**Figure 3.64**   *First baseline attempt—valleys descend below it*

If valleys do intersect the baseline, the integrator registers the first of the valleys as a baseline point and draws a baseline section from the start of the first peak to this valley. A second section is drawn from the valley to the end of the group (Figure 3.65). The situation is re-assessed to determine whether any valleys intersect the reconstruction. If none does, no further amendments are required and the peaks can be measured.

If intersecting valleys still exist, a new section of baseline is drawn from the first baseline valley to the next intersecting valley which is confirmed as a baseline point. A third section of baseline is drawn from the second 'baseline valley' to the end of the group, and this new section is inspected for intersecting valleys (Figure 3.66).

The process continues, moving forward in time, until a baseline is constructed which skirts round the lowest valleys. Perpendiculars are dropped from the other valleys to the completed baseline. This has been

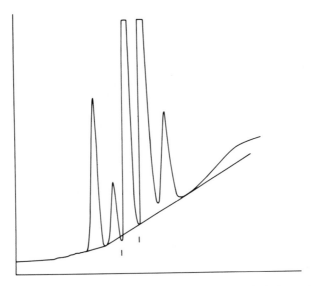

**Figure 3.65** *Second attempt; baseline updated to first valley which descended—but a valley still descends below second attempt*

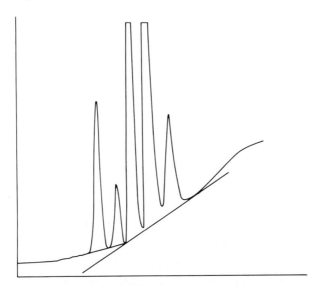

**Figure 3.66** *Third attempt; baseline updated to second descending valley. This time there are no remaining valleys which cut through the baseline*

likened to stretching a rubber band round the underside of the peaks touching the start and end points of the group and all the low valleys (Figure 3.67).

As valleys become confirmed baseline points, other valleys which might have been designated as baseline points in early constructions become redesignated as valleys by the updates.

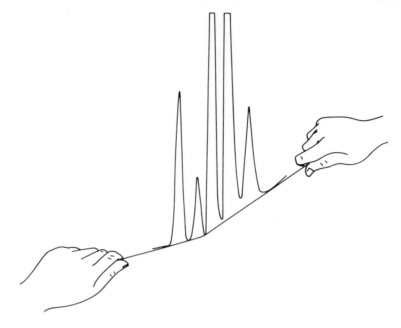

**Figure 3.67** *The Elastic Band Technique. Integrators fit a baseline under a group of peaks in a manner analogous to stretching an elastic band around the underside*

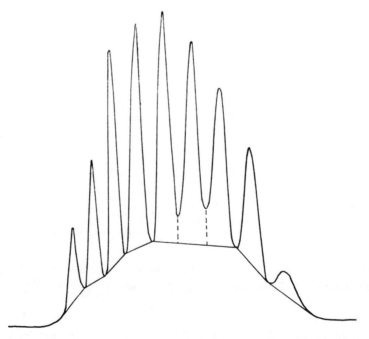

**Figure 3.68** *Valley–valley skim. It ignores the mass beneath the peaks*

**Valley–Valley Skim**

Some analysts have in the past preferred to process a complex, unresolved bunch of peaks by skimming a baseline around the valley bottoms rather than draw one baseline across the whole group and drop perpendiculars to it. This can be done by giving the baseline drift a high tolerance so that each valley (or at least some) comes within baseline range as the baseline updates itself to the preceding valley (Figure 3.68).

The use of this technique seems to be fading with improvements in WCOT technology.

# 7 Conclusions

It should be clear now that integrators are like any other tool—an excellent thing in the right hands. What they do best is measure peaks which are suitable for measuring, rapidly and without tedium. If these measurements are worth making then all subsequent calculations are worth noting and perhaps acting upon.

As long as integrators use perpendiculars and tangents and draw straight baselines beneath peaks, they are of use only in controlled circumstances, when the chromatography is good. They cannot improve bad chromatography: only the analyst can do that—and at the end of the day that's what he is paid for.

# 8 References

1. D. T. Sawyer and J. K. Barr, *Anal. Chem.*, 1962, **34**, 1213.
2. G. L. Feldman, M. Maude, and A. Windeler, *International Laboratory*, March/April 1971.
3. A. T. Leung, J. R. Hubbard, and L. A. Miller, *J. Chromatogr. Sci.*, 1976, **14**, 166.
4. B. Vandeginste, R. Essers, T. Bosman, J. Reijnen, and G. Kateman, *Anal. Chem.*, 1985, **57**, 971.
5. J. Novak, 'Advances in Chromatography', Vol. 11, Marcel Dekker, New York, 1973.
6. 'Modern Practice of Gas Chromatography', ed. R. L. Grob, 2nd Edn., Wiley Interscience, New York, 1985.
7. L. R. Snyder and J. J. Kirkland, 'Introduction to Modern Liquid Chromatography', 2nd Edn., Wiley Interscience, New York, 1979.
8. 'Quantitative Analysis using Chromatographic Techniques', ed. E. Katz, Separation Science Series, John Wiley and Sons, New York, 1987.
9. G. Guiochon and C. L. Guillemin, 'Quantitative Gas Chromatography', *J. Chromatogr. Library*, 1988, **42.**
10. G. B. Clayton, 'Data Converters', Macmillan Press, London, 1982.
11. H. H. Willard, L. L. Merritt Jr., J. A. Dean, and F. A. Settle Jr., 'Instrumental Methods of Analysis', 7th Edn., Wadsworth Pub. Co., California, 1988.
12. D. A. Skoog, 'Principles of Instrumental Analysis', 3rd Edn., Saunders College Pub., Philadelphia, Pa., 1985.
13. Burr Brown Application Note PDS 372 B, 'VFC32 V/F and F/V Converter', Tucson, Ariz., 1982.

14. Z. Hippe, A. Bierrowska, and T. Pietryga, *Anal. Chim. Acta*, 1980, **122**, 279.
15. A. H. Anderson, T. C. Gibb, and A. B. Littlewood, *Anal. Chem.*, 1970, **42**, 434.
16. J. Novak, K. Petrovic, and S. Wicar, *J. Chromatogr. Sci.*, 1971, **55**, 221.
17. S. N. Chesler and S. P. Cram, *Anal. Chem.* 1971, **43**, 1922.
18. M. Goedert and G. Guiochon, *Chromatographia*, 1973, **6**, 76.
19. D. T. Rossi, *J. Chromatogr. Sci.,* 1988, **26**, 101.
20. R. Bracewell, 'The Fourier Transform and its Applications', McGraw Hill, New York, 1965.
21. J. T. Tou, 'Digital and Sampled Data Control Systems', McGraw Hill, New York, 1959.
22. K. Kishimoto and S. Musha, *J. Chromatogr. Sci.*, 1971, **9**, 608.
23. J. D. Hettinger, J. R. Hubbard, J. M. Gill, and L. A. Miller, *J. Chromatogr. Sci.*, 1971, **9**, 710.
24. A. Fozard, J. J. Frances, and A. Wyatt, *Chromatographia*, 1972, **5**, 377.
25. P. C. Kelly and G. Horlick, *Anal. Chem.*, 1973, **45**, 518.
26. R. Q. Thompson, *J. Chem. Educ.*, 1985, **62**, 866.
27. A. Savitsky and M. J. E. Golay, *Anal. Chem.*, 1964, **36**, 1627.
28. D. L. Massart, B. G. M. Vandeginste, S. N. Deeming, Y. Michotte, and L. Kaufman, 'Chemometrics: a Textbook', Elsevier, Amsterdam, Chap. 15, 1988.
29. S. P. Cram, S. N. Chesler, and A. C. Brown, *J. Chromatogr.*, 1976, **126**, 279.
30. J. Steiner, Y. Termonia, and J. Deltour, *Anal. Chem.*, 1972, **44**, 1906.
31. M. H. J. van Rijswick, 'Philips Res. Report Sup.' Centrex Pub. Co., Eindhoven, No. 7, 1974.
32. A. W. Westerberg, *Anal. Chem.,* 1969, **41**, 1770.

# Subject Index

ASTM 42, 65
Accountability 8–9, 106
Accurate representation of solute
    profile 31–40
Additivity of detector signal 32
Adsorption isotherm 6, 40
Analog to digital conversion 107–110
Analysis parameters 94–101
  analysis duration 97
  baseline drift tolerance 96, 142–144,
    147–151
  minimum area/height threshold 97
  parameter over-ride 97–101
  peak width 95, 114–115
  slope sensitivity 95, 126, 128–129
  time to double 96
Analysis report 9, 93, 105, 106
Analysis reprocessing *see* Reprocessing
Analyst workload 9
Aperture time jitter 116
Area height ratio 5, 11, 12
Area% 63, 103
Assigning baseline beneath chromato-
    gram 151–155
Asymmetry 5, 17–19, 28, 59–63, 73
  and manual peak measurement 81–83
  *see also* Excess *and* Skew
Asymmetry ratio 6, 12, 28, 81
Auto ranging of A/D converters 110

Baseline 4, 14, 20, 45–49, 60, 64, 89, 91,
    96, 99–100, 126, 128, 131, 134,
    142–155
  definition of 45, 144
  drifting 43, 45–49, 142–144
Baseline construction 46, 59, 61, 78,
    133, 149, 152–155
Baseline signal level 4,127
Baseline subtraction 123

BASIC programming 93, 94, 103
Boxcar averaging 118

Calculations 90, 103–105
Calibration 77, 83, 105
  errors 77, 105
Chromatogram storage 93, 106
  reprocessing 93, 100, 117
  subtraction 122, 123
Chromatograph utilization 8
Coefficient of variation 29
Column efficiency *see* Efficiency
Column non-linearity 40
Column suitability 25
Computer simulation of peaks 18, 55,
    82
Condal-Bosch measurement 26, 29, 80
Cost per analysis 9, 86, 95
Counting squares 73
Curve fitting 58, 59, 92
Cutting and weighing 73

Data bunching 95, 112–115
Data sampling 95, 110–115, 127–131
  and peak width parameter 114
Data storage 93, 106, 112
Dead band 4, 71, 129
Detectors 31–40
  flame ionization 33, 35, 38, 64
  flow sensitive 34–37, 64
  mass sensitive 34–37, 64
  linearity 38–40, 64
  noise 33, 41–45
  overload 23, 25, 38, 39, 40, 64
  reduced operating range 48
  thermal conductivity 35, 38, 64
  three dimensional vi
  ultra-violet 34, 36, 64

Detector signal 32, 33, 69
  back-off control 48, 128
  zeroing 48, 72
Digital to analog conversion 33, 123
Digital chromatogram plotting 123
Digital measurement of peak areas
    107–141
Disk integrator 87–88
Distortion *see* Peak shape distortion
Distribution curve 9, 17, 20, 73
Dual slope integrating A/D conversion
    108
Dynamic equilibrium 6

EMG *see* Exponentially modified
    Gaussian
Efficiency 4–5, 19, 61
Elastic band technique 154
Electromechanical counters 87–90
  disadvantages 89
Error diagnostics 25, 137, 140
Error function 15, 18, 19
Errors created by noise 41, 43, 60
Errors in baseline construction 45–49,
    145–151
Errors in peak area measurement
    29–65, 78, 82–85, 145
  accuracy 29–31, 45, 65, 74, 82
  precision 29–31, 45, 65, 84
  random errors 30
  repeatability 31
  reproducibility 31
  systematic errors 30
Errors in programming integrators 115,
    129, 147
Errors of incomplete peak resolution
    49–59
Errors of peak asymmetry 59–63
Errors of perpendicular/tangent
    transition 52, 86
Errors of tangent skim 52
Event marks 91, 131, 132
Event program 95, 101
Excess 22–23, 45
Experimental validation 1, 4–7
Exponential function 16, 59, 61, 134
Exponentially modified Gaussian
    (EMG) function 17–19, 75, 81
  measurement of EMG peaks 28–29,
    81
  peak shape tests 19, 81
External standard 104
Extra column broadening 5, 22, 59

FDA 6

False peak starts 91, 98, 126, 144
Fastest measurable peak 131
Files 94
Flow rate 36
Forcing baseline detection 99, 146
Forcing tangent/perpendicular
    separation 52, 98, 138
Fractional peak area 13, 63
Fractional peak height 11, 63, 85
Fused groups 58, 137, 148

GLP 8
GPC 115
Gaussian function 9–17
  derivatives 10, 12, 125, 126
  infinite boundaries 13
  key dimensions 12
  parabolic maximum 16
  peak shape tests 12
Glossary of Terms xiii

Height 10–11
Height of fused peak 55
Higher statistical moments 24
History of integrators 87–93
Horizontal baseline projection 89, 100

Inflection points 12, 27, 76
Injection volume 8
Instrument control 95, 101
Integrate inhibit 98, 146
Integrators
  disk 87
  electromechanical 87–90
  electronic (TTL) 90–92
  microprocessor based 92
  'standard specification' 93–94
Internal standard 103
Inverting negative peaks 100

Keyboard indicators 91, 127, 131, 132

Laboratory information systems 93,
    106
Least mean squares *see* Linear
    regression
Limits of detection 32, 44, 53
Limits of integration 60, 65, 125,
    129–132, 146
Limit of quantitation 44
Linear regression 104, 119
Linearity *see* Detector linearity
Location of peaks 125–132

MDQ *see* Minimum detectable quantity

Manual peak area measurement 25–29,
  73–84
  advantages and disadvantages 85–86
  boundary methods 73–74
  Condal-Bosch area 26, 78
  counting squares 73
  cutting and weighing 73
  Foley variations 28–29, 81
  height × width at half height 25,
    77–78
  measurement of asymmetric
    (EMG) peaks 28–29
  peak area by triangulation 27,
    76–77
  planimetry 74
Maximum peak slope 13
Mean retention time 3, 20, 25
Measurement diagnostics 137, 140, 148
Measurement strategies 73–86
  counting squares 73
  cutting and weighing 73
  planimeters 74
Measurement speed 65, 75, 86, 126
Method development 5, 30, 94, 146
Microcomputers 92
  impact of microcomputer 92–93
Microprocessor based integrators 91
Minimum detectable quantity 39, 43
Minimum peak area 91
Mobile phase 9, 35
Mode retention time 3
Moments *see* Statistical moments
Moving window 118–122
Multilinear calibration 104
Multiple fused peaks 58, 135–137

NAMAS 8
Negative baseline dips 100, 149–151
Negative peaks 100
Noise 4, 41–45, 63, 116–122
  drift 43
  frequency 41
  long term noise 43, 97
  short term noise 41
Noise filtering 33, 64, 116–122
Noise test 95
Normal distribution function *see*
  Gaussian function
Normalization *see* Area%
Normalization of moments 20–23
Nyquist sampling frequency 111

Optimum filtering 120–122
Optimum peak shape 78, 84–85

Overlap and area measurement
  accuracy 52–58, 148
Overlapping peaks 32, 49–58, 61, 62,
  65, 77, 81, 92, 135–137
Overlapping peaks on sloping baselines,
  56–57

Peak area 1
  area and solute quantity 35–38, 63,
    103–105
  area, total 8
  asymmetry 5, 17–19, 63, 120, 121; *see
    also* Skew *and* Excess
  loss from base 14, 71
  loss from flanks 63
  loss from peak top 38, 64
Peak deconvolution vi, 55, 58–59, 92
Peak end 131–132
Peak height 1, 10, 33, 55, 75
  loss due to non-linearity 64
  measurement 55, 75, 76, 134
  area versus height 63–65, 75–76
Peak identity 3, 102
Peak resolution 6, 49–59
Peak shape distortion 32–34, 120, 121
Peak shape tests 12, 19, 81, 125, 126
Peak start 129–131
Peak width 4, 11, 33
Peak width parameter 95, 114, 115, 129
Pencil and rule methods 75–81
Performance measurement 8–9, 106
Perpendicular separation 49, 50–55, 92
Planimeter 74
Points of inflection 12, 121, 125
Polynomial curve fitting 104, 118–122
Post run processing 86, 137
Printer plotters 91
Processing speed *see* Measurement
  speed

Quadrants of a peak 125

Random errors 30
Recorders *see* Strip chart recorders
Relative retention times 102
Relative standard deviation 29
Relays 101
Repeatability 31
Reprocessing data 100, 115, 117
Reproducibility 31
Report distribution 9, 93, 106
Representative baseline 128
Resolution 6, 32, 49–59
Resolution of A/D converters 110, 116
Response factors 3, 103–105

Results validation 1, 4, 7, 8, 86
Retention time 3, 7, 126, 134
Retention time shift 46–48
Rounding errors 116

Sample size *see* Injection volume
Sampling frequency 107, 110–112, 120
    and bunch size 112
    and data processing algorithms 112
    and integrator manufacture 112
    and mains coupling 117
Savitsky–Golay 119–120, 125, 127, 135
Saturation *see* Detector saturation
Scaling factor 103
Shoulders 53, 138
Signal subtraction 122–123
Signal/noise ratio 43–44, 45, 65, 112,
    115, 116–121
Skew 22, 29, 45, 59–62
Slope sensitivity 95, 128–129, 138
Slope test 95
Smallest measurable peak 43
Small peak filtering 131
Smoothing 3, 116–120
Software 92
Solute identity 1, 3, 102
Solute profile 31–34, 85
Standard calculations 93, 103–105
Standard integrator specification 93–94
Stationary phase 9
Statistical moments (0th to 4th) 20–25
    higher moments 24
    measurement of peak moments 24
    practical disadvantages 25
Strip chart recorder 69–72, 87
    amplifier noise 71
    attenuator accuracy 72
    chart motor control 72
    dead band 71
    non-linear signal response 69
    pen head damping 70
    slow response time 69

Synchronization
    of instrument operations 101
    of peak sampling and peak start 115
System controller 94
System suitability 94
Systematic errors 30

Tangent separation 49, 138–141, 148
Tangent skim errors 55, 56–57, 140
Theoretical plates 19, 109
Time band and window 3, 102
Time constant 33, 34, 72
Time program 95, 97–101
Time to double 96, 129
Total peak area 8
Trapezoidal baseline correction 133,
    136, 137
Triangulation 27, 49, 53, 76
Truncation errors 116
TTL integrators 90

Unresolved peaks *see* Overlap *and*
    Fused groups
Updating parameters *see* Time program
    *and* Time to double
US Pharmacopoeia 6
Utilization of chromatographs 8

VFC *see* Voltage to frequency
    conversion
Validation 1, 4, 8, 86, 94
Valleys 56, 135, 149, 152
Valley shift 56
Valley–valley skim 155
Variance 11, 19, 21, 29, 45, 48, 83
Voltage to frequency conversion 90,
    107

WCOT peaks 32, 33, 70, 92, 137, 155